Praise for

GRUNT

"Nobody does weird science quite like [Roach], and this time, she takes on war. Though all her books look at the human body in extreme situations (sex! space! death!), this isn't simply a blood-drenched affair. Instead, Roach looks at the unexpected things that take place behind the scenes."
—*Wired*

"Roach . . . applies her tenacious reporting and quirky point of view to efforts by scientists to conquer some of the soldier's worst enemies."
—*Seattle Times*

"Extremely likable . . . and quick with a quip. . . . [Roach's] skill is to draw out the good humor and honesty of both the subjects and practitioners of these white arts among the dark arts of war."
—*San Francisco Chronicle*

"Covering these topics and more, Roach has done a fascinating job of portraying unexpected, creative sides of military science."
—*New York Post*

"Science writer Mary Roach . . . dives into the world of military science. Roach learns just how much research goes into every aspect of preparing for war, from figuring out how to deal with diarrhea in the field to designing a camouflage pattern that doesn't get men killed to finding ways for service members on submarines to get some shut-eye."
—*Mental Floss*

"[Roach] never fails to notice a pun, or an irony, or a dry observation. . . . The unflagging enthusiasm in her books, the raw happiness that bounces off the pages, isn't the sort of thing that can be faked."

—*Seattle Review of Books*

"Roach uses humor and a lucid narrative to describe the human body's physiological reactions to extreme heat and cold, noise, fatigue, smell, combat-weight burdens, sleep deprivation, and 'ill-timed gastrointestinal urgency' of battlefield diarrhea. . . . She also tells of scientists and engineers working to mitigate the adverse effects of these conditions, as well as new training, procedures, and inventions to reduce impacts on service members."

—*Military Officer*

"Roach hilariously and delightfully teaches us what happens when you run out of space on a nuclear submarine, . . . debates the need for bombproof underwear, and volunteers to be shot by a paintball-gun-toting Marine at 70 feet."

—*San Francisco Chronicle* "Top Shelf"

"Roach skillfully moves from topic to topic throughout the book, creating a connected narrative. Through Roach's energetic writing, the reader feels like they are along for the ride, discovering the interesting science behind the military with the author." —*Times-News*

"In the amusing and thought-provoking *Grunt*, [Roach] applies her special mix of humor and insatiable curiosity to the work of scientists of the American military who 'run along in the wake of combat, lab coats flapping.' " —*Columbus Dispatch*

"A treasure trove of weird, obscure military science trivia; every chapter has at least one line that will make you sit up and say, 'Huh. Who knew?' "

—*Vox*

"For everything other than the killing, it's a masterful, digestible look at the science of war. . . . In telling the tale, Roach details history from nuclear tests and Spanish-American War fashion trends, as well as modern adaptations to unisex uniforms. The chapters are rich with information, but they never feel weighed down by their density. There is a flow to Roach's work: from uniforms *Grunt* moves on to surviving bomb blasts in vehicles, then to the challenge of preserving hearing while listening for danger on a battlefield."

—*Popular Science*

ALSO BY MARY ROACH

Gulp: Adventures on the Alimentary Canal

Packing for Mars: The Curious Science of Life in the Void

Bonk: The Curious Coupling of Science and Sex

Spook: Science Tackles the Afterlife

Stiff: The Curious Lives of Human Cadavers

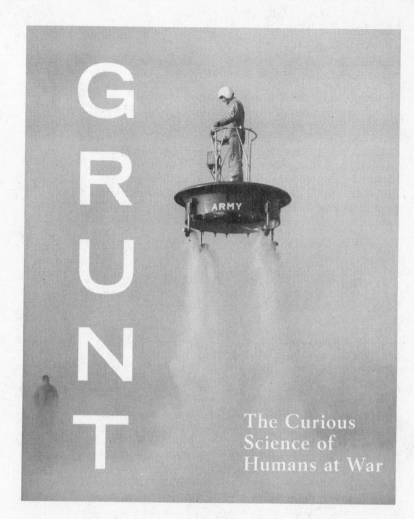

G R U N T

The Curious Science of Humans at War

Mary Roach

W. W. NORTON & COMPANY | *Independent Publishers Since 1923* | New York | London

Copyright © 2016 by Mary Roach

All rights reserved
Printed in the United States of America
First published as a Norton paperback 2017

Photograph credits: Frontispiece: Nat Farbman / The LIFE Picture Collection / Getty Images. Chapter 1: U.S. Army photo. Chapter 2: Library of Congress. Chapter 3: Museum Waalsdorp. Chapter 4: Gabriel Benzur / The LIFE Images Collection / Getty Images. Chapter 5: Science Photo Library / Alamy Stock Photo. Chapter 6: Strategic Operations, Inc. Chapter 7: GAMMA / Gamma-Keystone via Getty Images. Chapter 8: Photographer's Choice / Getty Images. Chapter 9: National Museum of Health and Medicine, https://creativecommons.org/licenses/by/2.0/. Chapter 10: AP Photo. Chapter 11: Nature Picture Library / Alamy Stock Photo. Chapter 12: Hulton-Deutsch Collection / Corbis. Chapter 13: Buzz Pictures / Alamy Stock Photo. Chapter 14: Derek Hudson / Getty Images.

For information about permission to reproduce selections from this book, write to Permissions, W. W. Norton & Company, Inc., 500 Fifth Avenue, New York, NY 10110

For information about special discounts for bulk purchases, please contact W. W. Norton Special Sales at specialsales@wwnorton.com or 800-233-4830

Manufacturing by LSC Communications, Harrisonburg
Book design by Ellen Cipriano
Production manager: Anna Oler

Library of Congress Cataloging-in-Publication Data

Names: Roach, Mary.
Title: Grunt : the curious science of humans at war / Mary Roach.
Description: First edition. | New York : W. W. Norton & Company, Inc., 2016. | Includes bibliographical references.
Identifiers: LCCN 2016008754 | ISBN 9780393245448 (hardcover)
Subjects: LCSH: Military art and science—United States—Technological innovations. | Military art and science—Technological innovations—United States—History—21st century. | Military research—United States.
Classification: LCC U43.U4 R63 2016b | DDC 355/.070973—dc23
LC record available at https://lccn.loc.gov/2016008754

ISBN 978-0-393-35437-9 pbk.

W. W. Norton & Company, Inc.
500 Fifth Avenue, New York, N.Y. 10110
www.wwnorton.com

W. W. Norton & Company Ltd.
15 Carlisle Street, London W1D 3BS

1 2 3 4 5 6 7 8 9 0

In memory of William S. Rachles

CONTENTS

BY WAY OF INTRODUCTION   13

1   SECOND SKIN   18
What to wear to war

2   BOOM BOX   40
Automotive safety for people who drive on bombs

3   FIGHTING BY EAR   56
The conundrum of military noise

4   BELOW THE BELT   72
The cruelest shot of all

5   IT COULD GET WEIRD   88
A salute to genital transplants

6   CARNAGE UNDER FIRE   104
How do combat medics cope?

7   SWEATING BULLETS   124
The war on heat

8   LEAKY SEALS   142
Diarrhea as a threat to national security

9   THE MAGGOT PARADOX   164
Flies on the battlefield, for better and worse

10   WHAT DOESN'T KILL YOU
WILL MAKE YOU REEK   184
A brief history of stink bombs

11   OLD CHUM   202
How to make and test shark repellent

12   THAT SINKING FEELING   222
When things go wrong under the sea

13   UP AND UNDER   242
A submarine tries to sleep

14   FEEDBACK FROM THE FALLEN   264
How the dead help the living stay that way

ACKNOWLEDGMENTS   273

BIBLIOGRAPHY   277

GRUNT

By Way of Introduction

THE CHICKEN GUN HAS a sixty-foot barrel, putting it solidly in the class of an artillery piece. While a four-pound chicken hurtling in excess of 400 miles per hour is a lethal projectile, the intent is not to kill. On the contrary, the chicken gun was designed to keep people alive. The carcasses are fired at jets, standing empty or occupied by "simulated crew," to test their ability to withstand what the Air Force and the aviation industry, with signature clipped machismo, call *birdstrike*. The chickens are stunt doubles for geese, gulls, ducks, and the rest of the collective bird mass that three thousand or so times a year collide with Air Force jets, costing $50 million to $80 million in damage and, once every few years, the lives of the people on board.

As a bird to represent all birds, the chicken is an unusual choice, in that it doesn't fly. It does not strike a jet in the manner in which a mallard or goose strikes a jet—wings outstretched, legs trailing long. It hits it like a flung grocery item. Domestic chickens are, furthermore, denser than birds that fly or float around in wetlands.

At 0.92 grams per centimeter cubed, the average body density of *Gallus gallus domesticus* is a third again that of a herring gull or a Canada goose. Nonetheless, the chicken was the standard "material" approved by the US Department of Defense for testing jet canopy windows. Not only are chickens easier to obtain and standardize, but they serve as a sort of worst-case scenario.

Except when they don't. A small, compact bird like a starling can pierce a canopy windscreen like a bullet, and apparently does so often enough that someone saw fit to launch some jargon (the "feathered bullet phenomenon"). Would it be simpler to just keep birds away from runways? You'd think. But birds habituate. They quickly adjust to whatever predator sound or alarm call you broadcast or minor explosives you set off, just "singing or calling more loudly"* and going about their lives as they always have.

Enter Malcolm Kelley and the Bird Aircraft Strike Hazard (BASH) team of the United States Air Force. Kelley and his team took a cross-disciplinary approach. Engineering, say hello to biology. Ornithology, meet statistics. Let's break this down, they said. Let's start with turkey vultures. Though implicated in only 1 percent of Air Force birdstrikes, the weighty raptors are, by one accounting, responsible for 40 percent of the damage. Kelley and the team attached transmitters to eight of them, tracked their flight habits and patterns, and combined this with other data to create a Bird Avoidance Model (BAM) that would enable flight schedulers to avoid high-risk times and air space. A simple "improvement in

---

* I quote the paper "What Can Birds Hear?" The author, Robert Beason, notes that acoustic signals work best when "reinforced with activities that produce death or a painful experience . . ." He meant for some members of the flock, whereupon the rest would presumably take note. As would animal rights activists, producing a painful experience for public affairs staff.

Turkey Vulture understanding" had, Kelley projected, the potential to save the Air Force $5 million per year, as well as the lives of unknown numbers of pilots (and turkey vultures).

Sifting through the data, Kelley noticed that when the frequency range of a jet engine sound overlapped with the frequency range of a species' distress call, the likelihood of birdstrike appeared to be lower. "Are we talking to the birds without realizing it?" he wrote in a 1998 paper. Might there be a way to build on this? One problem, he knew, is that both birds and planes take off facing into the wind. Thus the former often do not see the latter bearing down on them from behind. It was Kelley's idea to add a meaningful signal to an aircraft's radar beam, something that would alert birds to the danger sooner, so they'd have time to react and get out of the way.

This is the sort of story that drew me to military science—the quiet, esoteric battles with less considered adversaries: exhaustion, shock, bacteria, panic, ducks. Surprising, occasionally game-changing things happen when flights of unorthodox thinking* collide with large, abiding research budgets. People tend to think of military science as strategy and weapons—fighting, bombing, advancing. All that I leave to the memoir writers and historians. I'm interested in the parts no one makes movies about—not the killing but the keeping alive. Even if what people are being kept alive for is fighting and taking other lives. Let's not let that get in the way. This book is a salute to the scientists and the surgeons, running along in the wake

---

* Kelley's furthest foray outside the box came at a 1994 Wright Laboratory brainstorming session on nonlethal weapons. In the category of "chemicals to spray on enemy positions," he came up with "strong aphrodisiacs." Was the idea to develop a compound that would generate feelings of love for the enemy? "No," Kelley said. "The idea was to ruin their morale because they're worried their buddy is going to come in their foxhole and make fond advances." And come in their foxhole.

of combat, lab coats flapping. Building safer tanks, waging war on filth flies. Understanding turkey vultures.

THE CHICKEN gun is most of what I have to say about guns. If you're wanting to read about the science of military armaments, this is not the book you're wanting to read. Likewise, this is no *Zero Dark Thirty.* I talk to Special Operations men—Navy SEALs and Army Rangers—but not about battling insurgents. Here they're battling extreme heat, cataclysmic noise, ill-timed gastrointestinal urgency.

For every general and Medal of Honor winner, there are a hundred military scientists whose names you'll never hear. The work I write about represents a fraction of a percent of all that goes on. I have omitted whole disciplines of worthy endeavor. There is no chapter on countermeasures for post-traumatic stress disorder, for example, not because PTSD doesn't deserve coverage but because it has had so much, and so much of it is so very good. These books and articles aim the spotlight where it belongs. I am not, by trade or character, a spotlight operator. I'm the goober with a flashlight, stumbling into corners and crannies, not looking for anything specific but knowing when I've found it.

Courage doesn't always carry a gun or a flag or even a stretcher. Courage is Navy flight surgeon Angus Rupert, flying blindfolded and upside down to test a vibrating suit that lets pilots fly by feel should they become blinded or disoriented. It's Lieutenant Commander Charles "Swede" Momsen, saluting onlookers as he's lowered into the Potomac to test the first-ever submarine escape lung, or Captain Herschel Flowers of the Army Medical Research Laboratory, injecting himself with cobra venom to test the possibility of building immunity. Sometimes courage is nothing more than a willingness to think differently than those around you. In a culture of conformity, that's braver than it sounds. Courage is World War I medic William Baer, saving limbs and lives by letting maggots debride wounds. It's Dr. Herman Muller, volunteering to inject himself with cadaver blood to test the safety of transfusions from the dead to the wounded, a practice carried out on the battlefields of the Spanish-American War.

Heroism doesn't always happen in a burst of glory. Sometimes small triumphs and large hearts change the course of history. Sometimes a chicken can save a man's life.

1

Second Skin

What to wear to war

A N ARMY CHAPLAIN IS a man of the cloth, but which cloth? If he's traveling with a field artillery unit, he is a man of moderately flame-resistant, insect-repellent rayon-nylon with 25 percent Kevlar for added durability. Inside a tank, he's a man of Nomex—highly flame-resistant but too expensive for everyday wear. In the relative safety of a large base, the chaplain is a man of 50/50 nylon-cotton—the cloth of the basic Army Combat Uniform, as well as the camouflage-print vestments that hang in the chaplain's office here at Natick Labs.

The full and formal title of the complex of labs known casually as "Natick" is US Army Natick Soldier Research, Development and Engineering Center. Everything a soldier wears, eats, sleeps on, or lives in is developed or at least tested here. That has included, over the years and through the various incarnations of this place: self-heating parkas, freeze-dried coffee, Gore-Tex, Kevlar, permethrin, concealable body armor, synthetic goose down, recombinant spider silk, restructured steaks, radappertized ham, and an emergency ration chocolate bar with a dash of kerosene to prevent ad libitum snacking. Natick chaplains, for their part, have devised portable confessionals, containerized chapels, and extended shelf life* communion wafers.

It's a balmy 68 degrees at Natick this afternoon. It may, at the same time, be 70 below zero with horizontally blown snow or 110 in the shade, depending on what's being tested over in the Doriot. The Doriot Climatic Chambers were the centerpiece of the complex when it opened in 1954. Never again would troops be sent to the Aleutian Islands with seeping, uninsulated boots or to equatorial jungles with no mildew-proofing on their tents. Soldiers fight on their stomachs, but also on their toes and fingers and a decent night's sleep.

These days, the snow and rain machines are rented out to L.L. Bean or Cabela's as often as they're used to test military outerwear. Repelling the elements is the least of what the US Army needs its uniforms to do. If possible, the army would like to dress its men and women in uniforms that protect them against all that modern warfare has to throw at them: flames, explosives, bullets, lasers, bomb-blasted dirt, blister agents, anthrax, sand fleas. They would like these same uniforms to keep soldiers cool and dry in extreme heat, to stand up to the ruthless rigors of the Army field laundry,

---

* Natick Labs and precursor the Quartermaster Subsistence Research Laboratory have extended shelf lives to near immortality. They currently make a sandwich that keeps for three years. Meat, in particular, has come a long way since the Revolutionary and Civil wars, when beef came fresh off cattle driven alongside the troops. During World War II, the aptly named subsistence lab developed partly hydrogenated, no-melt "war lard" and heavily salted and cured, extra-dry "war hams" that kept for six months without refrigeration and earned the not exactly over-the-moon descriptors "palatable and satisfactory." I quote there the July–August 1943 *Breeder's Gazette*, sister publication of the *Poultry Tribune*, a newspaper either about or for barnyard fowl.

to feel good against the skin, to look smart, and to come in under budget. It might be easier to resolve the conflicts in the Middle East.

L ET US begin at Building 110, which is what everyone calls it. Officially it was christened the Ouellette* Thermal Test Facility, lending a flirtatious French flair to lethal explosions and disfiguring burns. The head textile technologist is a slim, classy, fiftyish woman of fine-grained good looks, dressed today in a cream-colored cable-knit wool tunic. I took her to be the Ouellette, and then she opened her mouth to speak and a hammered-flat Boston accent flew out and slammed into my ear. She is an Auerbach, Margaret Auerbach, but around 110 she's just Peggy, or "flame goddess."

When someone in industry thinks they've built a better flame-resistant fabric, a sample comes to Auerbach for testing. Some people submit swatches; others optimistically ship off whole bolts. Their hopes may be undone by a single strand of thread. "To see what our guys might be inhaling," Auerbach heats a few centimeters of thread to around 1500 degrees Fahrenheit. The fumes produced by this are identified by gas chromatography. Flame-resistant textiles—some, anyway—work via heat-released chemicals. Auerbach needs to be sure the chemicals aren't more dangerous than the flames themselves.

Once it's established that the textile is nontoxic, Auerbach sets about testing its flame-stopping mettle. This is done in part with a Big Scary Laser (as the sticker on its side reads). Auerbach places a swatch in the laser's sights. And here is the best part: To activate this laser, you push a *giant red button*. The beam is calibrated to deliver a

---

* Misspelled as "Uoellette" on the Natick Building Inventory, and "Oullette" on the sign outside the building. Somebody burned for that one.

scaled-down burst of energy representative of an insurgent's bomb—a teacup IED. A sensor behind the swatch measures the heat passing through, yielding a figure for how much protection the fabric provides and what degree burn would result.

Auerbach switches on a vacuum pump that sucks the swatch tight against the sensor. This is done to approximate an explosion's pressure wave—the dense pileup of accelerated air that can knock a person flat. More subtly, it forces clothing flush against the skin, which can heighten the heat transfer and worsen the burn. One of the winning attributes of Defender M, the textile of the current Flame Resistant Army Combat Uniform, or FR ACU ("the guys call it 'frack you'"), is that it balloons away from the body as it burns.

The downside to Defender M has been that it tears easily. (They're working on this.) The same thing that keeps it comfortable in hot weather also makes it weaker; it's mostly rayon, which draws moisture but has low "wet strength." If a garment tears open in the chaos of an explosion, now the protective thermal barrier is gone. Now you're toast. The manufacturer throws a little Kevlar in, but it still isn't as strong as Nomex, a fiber often used for firefighter uniforms. Nomex also has superior flame resistance: It buys you at least five seconds before your clothes ignite.

Auerbach explains that this is especially important for crews inside tanks and aircraft. "Where they can't roll, drop, and . . ." She rewinds. "Drop, stop . . . what is it?"

"Stop, drop, and roll?"

"Thank you."

Why not make all army uniforms out of Nomex? Poor moisture management. Not the best choice for troops running around sweating in the Middle East. And Nomex is expensive. And difficult to print with camouflage.

This is how it goes with protective textiles: Everything is a trade-off. Everything is a *problem*. Even the color. Darker colors reflect less heat; they absorb and transfer more of it to the skin. Auerbach goes across the lab to get a swatch of camouflage print cloth. She points to a black area. "You can see this has a pucker where it was absorbing more heat."

"It has a what?" I heard her, but I need to hear her say *pucka* again. The fabulous Boston accent.

I would have guessed the military to be a fan of polyester: strong, cheap, doesn't ignite. The problem is that it melts and, like wax and other melted items, it drips and sticks to nearby surfaces, thereby prolonging the contact time and worsening the burn. What you really don't want to be wearing inside a burning army tank is polyester tights.*

To determine what degree of injury the heat would produce, Auerbach runs the reading from the sensor behind the cloth through a burn prediction model—in this case, one developed after World War II by original flame goddess Alice Stoll. Stoll did burn research for the Navy. To work out first- and second-degree burn models, she gamely volunteered the skin of her own forearm. You may excuse her for letting someone else help out with the third-degree burn curve. Anesthetized animals were recruited for this—rats, mostly, and pigs. Pig skin reflects and absorbs heat in a manner more like our own than that of any other commonly available animal. The pig as a species deserves a Purple Heart, or maybe Pink.

---

* I went on the National Electronic Injury Surveillance System to find you a figure for the number of burns caused each year by pantyhose. Alas, NEISS doesn't break clothing injuries down by specific garment. I skimmed "thermal burns, daywear" until I ran out of patience, somewhere around the thirty-seven-year-old man who tried to iron his pants while wearing them.

What Stoll learned: When flesh reaches 111 degrees Fahrenheit, it starts to burn. The Stoll burn prediction model is a sort of mathematical meat thermometer. The heat of the meat and how deeply into the skin that heat penetrates are the critical factors that determine the degree of the burn. A brief exposure to flame or high heat cooks, if anything, just the outer layer, creating a first-degree burn or, to continue our culinary analogy, lightly seared ahi tuna. A longer exposure to the same heat cooks the inner layers, too. Now you have a second- or third-degree burn, or a medium-rare steak.

Even without a flame, clothing can catch fire. The auto-ignition temperature for cotton, for instance, is around 700 degrees Fahrenheit. Exposure time is key. The heat pulse from a nuclear blast is extremely hot, but it's traveling at the speed of light. Might it pass too quickly to ignite a man's uniform? Natick's early precursor, Quartermaster Research and Development, actually looked into this.

Operation Upshot-Knothole was a series of eleven experimental nuclear detonations at the Nevada Proving Grounds in the 1950s. The Upshot-Knothole scientists were mainly interested in the blastworthiness of building materials and tanks and bomb shelters, but they agreed to let the uniform guys truck over some pigs. Anesthetized Chester White swine, 111 in total, were outfitted in specially designed animal "ensembles" sewn from different fabric combinations—some flame-resistant and some not—and secured at increasing intervals from the blast.

Flame-resistant cool-weather uniforms with a layer of wool outperformed a series of thinner flame-resistant hot-weather uniforms—whose developers had surely, by "hot weather," not had in mind the extreme swelter of nuclear blast. The researchers marveled to note "a complete lack of any qualitative evidence of thermal injury to the fabric-protected skin of animals dead on recovery at the [1,850-foot] station." I don't wish to be an upshot-knothole, but who worries about burns on subjects close enough to a nuclear explosion that they are, as the report succinctly terms it, "blown apart"? Despite the clanging absurdity of the scenario, it was a memorable demonstration of the importance of exposure time. With the fast-traveling heat from a bomb—including a more survivable one like an IED—a few seconds of flame resistance can make all the difference.

The wool helped, too, because hair is naturally flame-resistant. Natick has, of late, been looking into a return to natural fibers like silk and wool. Not only is wool flame-resistant and nonmelting, it wicks moisture away from the body. Auerbach says she has seen some very nice, soft, flame-resistant cool-weather sheep's wool underwear. The hairs have to be descaled so the wool isn't itchy, and the garments need to be treated to keep them from shrinking, and both these processes add to the cost. As does the Berry Amendment, which gives preference to domestic suppliers of military gear. The Berry is additionally problematic in this case in that—despite the breathless, eager assurances of the American Sheep Farmer's Industry—there may not be enough sheep in all of America to fill the bill.

So let's say your new textile is comfortable and affordable. The flame resistance plays well with the insect-repellent treatment and the antimicrobial stink-proofing. Now what? Now you bring some over to the Textile Performance Testing Facility. You run it through the Nu-Martindale Abrasion and Pilling Tester to get a feel for how quickly the treatments will succumb to soldierly abuse. You subject it to a couple dozen wash and dry cycles. Laundering removes not only grime but also, bit by bit, the chemicals with which a cloth or fiber has been treated. When I visited the textile testing facility, a man named Steve was waiting for some pants to get through an accelerated wash. One wash in the Launder-Ometer equals five normal washes, he told me.

"That's something," I said.

"Yup." He stuck out his lower lip in a contemplative way. "Steel balls bang against the fabric."

If only the minds of Natick could invent a fabric that didn't need laundering. If everything splashed, smeared, or spilled on a uniform just beaded up and rolled off, if uniforms could be cleaned with a quick spray of water, think how much longer they'd last. And how much safer they'd be in the event chemical weapons rained down on them.

The minds of Natick are on it. Over in the liquid repellency evaluation lab, they're putting to the test a new "super-shedding" fabric treatment technology. Escorting me to a demo will be Natick's calm, likable public affairs officer, David Accetta. We meet up in his office, one side of which is piled with boxes from a recent move. A wall calendar features dog breeds. September is a large white poodle. Accetta was most recently deployed to Bagram Airfield in Afghanistan, where he spent his days writing press releases about the Army's humanitarian efforts. His superiors would ask him why the stories rarely got any play. "They didn't get it. It's not news." He relates this with no trace of anger. There are many irritating things about Accetta's job, but he never sounds irritated. He takes everything in stride, which is a bad cliché to use for him, because he's not a striding sort of guy. He's more of a moseyer. He has long eyelashes and a slow way of blinking. I almost wrote *doll-like* there, but the adjective seems out of place with the rest of Accetta's face, which is crossed by a thin, rakish scar that begins at one temple and curves down and around his cheek. I don't ask about it, preferring to supply my own made-up narrative of flashing sabers and staircase choreography.

We are early, so we take a walk along Lake Cochituate, which forms a property line for part of the Natick grounds. Sunlight is scattered on a low chop. Water from the lake, a deep blue-green in today's light, was at one time used to make Black Label lager. Natick activities pretty much put a stop to that. For a Superfund site, the grounds are quite pretty, with gazebos and meandering footpaths. Cylindrical gray-white Canada goose droppings add to the parklike atmosphere. It took a while to realize what these were, because I didn't see any geese. It's fall. Maybe they just flew south.

Accetta and I stop to watch an officer addressing a group of HRVs: human research volunteers—arms and feet and heads to go inside the parkas and boots and helmets. They are soldiers deployed to taste rations, sleep in new sleeping bags: test, report back, test something else. A temporary duty assignment at Natick is not necessarily a soft gig. I saw a photograph, from the sixties, of a group of soldiers in raincoats and waterproof pants, heads bent, hoods dripping, walking in circles under a simulated downpour. Apparently this went on for hours.

The volunteers, ten or so, stand in a row in the parking lot outside their barracks. A car backs out of a parking slot behind them. The soldiers take three steps forward, in formation, and one step up, onto the curb. When the car pulls away, they step backward and down. Anytime they walk someplace in a group of four or more, Accetta says, they have to be in formation. Like geese flying south.

.　　　.　　　.

THE DEMONSTRATION begins with the farting sound of a squeezable mustard bottle. A line of glistening yellow joins the duns and drab olives of a square of camouflage fabric. The cloth is clipped to a sloping board to foster roll-off. This being a roll-off test. As a cameraman and a small crowd look on, the line of mustard creeps down the cloth, holding its shape perfectly. A young chemical engineer, Natalie Pomerantz, directs onlookers' attention to the terrain across which the condiment has just traveled. "No residue trail!"

Ketchup follows, then coffee, and milk, as though the owner of the uniform had engaged the enemy in a food fight. Everything rinses clean with water. Natalie invites me to feel the underside of the cloth, and I do. It is completely dry.

Natalie started with the easy ones. Liquids that are mostly water have high surface tension. That means that the molecules prefer to bond more strongly to each other than to most of the things you might spill them on. A liquid with lower surface tension, like alcohol, won't bead up on a fabric the way water will; it soaks right in. A bead of water is a molecular huddle, a withdrawal inward, a refusal to join hands with strangers. Confronted with air, the surface of water pulls together powerfully enough to form a weak skin. The insect kingdom has water striders, but no gin striders. At the far high end of the surface-tension spectrum is mercury. Mercury beads up and rolls off pretty much any surface you drop it on without leaving a trace.* One of mercury's qualifications for old-timey thermometerdom—along with staying liquid in extremes of cold and heat—is that it doesn't

---

* Including the skin of children whose parents, like mine, collected the mercury from broken thermometers and put it in a plastic margarine tub for them to play with. The sixties sure were different.

wet the inside of the glass. No residue trail! So you can clearly read the temperature.

Many of the things the military tends to spurt and dribble on itself—motor oil, aviation fuel, hexane—have far lower surface tension than water. Natalie drips motor oil on another square of cloth. She picks up a cup of water and launches the contents like an outraged dinner date. Again, the mess rinses away with no trace remaining. "That was the money shot," a colleague says.

Natalie is nodding. Beaming, even. "This is like a day out in the sun for us." She's joking, but it's true in a way, in a good way. The delight she takes from science is an effervescence, something sparkly and hard to hold back. We should all love our jobs this much.

The super-repellent coating takes its inspiration from the leaves of the water lily. The surface of a lily pad, viewed under an electron microscope, is a carpet of tiny nubs, each covered with yet-smaller nano-nubs of waxlike crystals. (Paraffin wax is itself an effective fabric waterproofer, but it's too flammable for the military to use.) The tiny nubs and peaks reduce the contact and interaction between the cloth and whatever liquid spills on it. The coating also makes the surface more energetically stable, further discouraging interactions between the textile and the glop.

Though the publicity for this "super-shedding" coating has focused on its grime-shunning properties—"self-cleaning underwear" is how Accetta reeled me in—the more important application is protection from chemical and biological weapons. The first garments to feature the new technology will be an outer shell and pants: a chem/bio suit. Garments like this include a layer of activated carbon (also known as activated charcoal) to bind up noxious organic materials. The super-repellency means this layer can be thinner; if 95 percent of what hits a garment rolls right off it, that means far fewer activated carbon receptor sites are needed to bind up the poisons. That's good, because a garment with a thick layer of activated carbon is hot and uncomfortable. It's like wearing an air filter. With protective clothing, comfort is paramount. If it's uncomfortable, troops will be tempted to flout safety regulations and stuff it under their seat.

Likewise if they hate how it makes them look. "With protective gear especially, it's key that you design something that's kind of sleek and cool, because otherwise they're not going to want to wear it." That is a quote from Annette LaFleur, the US Army's top staff fashion designer and the next line item on "MARY ROACH SCHEDULE as of 1400 hrs 20 September 2013."

A ZIPPER IS a problem for a sniper. Here is a man who may spend an entire afternoon on his belly, sliding around on rubble and earth. If his top closes with a zipper, sand and muck will be ground into the spaces between the teeth, and soon it will jam. A zipper is, furthermore, uncomfortable to lie on. As are buttons. A "hook-and-loop fastener" like Velcro is a less protuberant front-closure option, but it's noisy. I have heard stories of Special Operations guys whose Velcro put them in danger by revealing their position. A stealthier model is on Natick's hook-and-loop fastener agenda.*

The designer of the new sniper base suit—Annette LaFleur, as pretty as the name suggests—is showing us how she got around this by making the top a side-closure garment. She indicates a dress form,

---

* Is it possible for an organization to have enough hook-and-loop fastener concerns to merit a whole agenda? When that organization is the US Army, it is. The US Army has enough hook-and-loop fastener concerns to merit an entire Hook and Loop Task Group (a sub-subcommittee of the Combat Clothing Utility Subcommittee).

standing in one of the work bays in Natick's Design, Pattern, and Prototype Facility. The headless dress form, a staple of the fashion industry, takes on an uncomfortable poignancy when the garment under construction is designed for war. Our sniper looks like someone else's sniper got him first.

LaFleur takes the fabric between her fingers. "This is a coated Cordura we went to." The bolts of fabric may be olive drab and the sewing machines built for Kevlar, but this is still a design studio, and LaFleur's language reflects this. (She described flame-resistant uniforms as one of the "things that are hot right now.") However, LaFleur didn't go to Cordura because it's in style. She chose it because it's durable and flame-resistant, and because the coated backing keeps moisture from seeping through. And that's important if you're lying someplace damp waiting to kill someone. Even the sleek, uncluttered lines of the sniper top are functional, a result of the side closure and of an earlier decision to move the pockets from the front to the sleeves for easier access. (The ensemble will stop looking sleek should elements of the accessory kit be added. Snipers can customize the back, pants, and helmet by tying in, say, jute strands to blend in with brush or grassy terrain. LaFleur, but probably not many snipers, compares this to macramé.)

LaFleur points to a cloth flap called a button placket, which covers the buttons so they don't get chipped. She needn't have worried. US military specifications for buttons include a minimum compressive strength, tested by placing the item between flat blocks of steel and bearing down until "the first audible sound of cracking." Federal button inspectors display a medieval zeal for their work. Other inspection methods include pressing a hot iron to the button's back, boiling it in water, and pulling the shank until it separates from the body.

US government button specifications run to twenty-two pages. This fact on its own yields a sense of what it is like to design garments for the Army. Although the Army requires its clothing designers to have a fashion design degree, fashion—in the sense of individuality expressed through appearance—is the opposite of an Army requirement. It's a violation of policy. US Army Appearance and Grooming Policies prohibit anything to which the adjectives "extreme, eccentric or faddish" might apply. The United States Army does not abide: unbalanced or lopsided hairstyles, "barrettes with butterflies," "large, lacy scrunchies," teased hair that rises more than 3½ inches from the scalp, hair that is dyed green, purple, blue, or "bright (fire-engine) red," Mohawks, dreadlocks, slanted or curved parts, flared sideburns, tapered sideburns, individual sideburn hairs that exceed 1/8 of an inch when fully extended, goatees, beards of any kind, Fu Manchus, and mustaches that cover any part of the upper lip or "extend sideways beyond a vertical line drawn upward from the corners of the mouth."*

The driving aesthetic of military style is uniformity. Whence the word *uniform*. From first inspection to Arlington National Cemetery, soldiers look like those around them: same hat, same boots, identical white grave marker. They are discouraged from looking unique, because that would encourage them to feel unique, to feel like an individual. The problem with individuals is that they think for themselves and *of* themselves, rather than for and of their unit.

---

* Defense Department facial hair policy is so complex that the Army saw fit to create visual aids for identifying Unauthorized Male Mustaches and Unauthorized Male Sideburns. Grids are overlaid on drawings of faces with labeled letter points, A, B, C, D. No such aid or information is provided for unauthorized female or transgender facial hair, and I encourage all such enlistees to begin growing their Fu Manchu and muttonchops now.

They're the lone goldfish on the old Pepperidge Farm bags, swimming the other way. They're a problem.

"You're more of an engineer than a designer," says LaFleur of her work. She got her start designing swimwear. It is a more logical transition than it might at first seem. A bathing suit requires expertise with high-performance active-wear fabrics and an understanding of the specialized activity they're needed for. Ditto, say, a concealable body armor vest. LaFleur's colleague Dalila Fernandez came to Natick from the now-defunct Priscilla of Boston, purveyors of high-end wedding dresses. Same thing here: A wedding gown entails multilayering of expensive specialty fabrics for an outfit whose useful lifespan may come and go in a single afternoon. Much like a bomb suit. Form follows function—although admittedly more so here than in most studios. Only a military clothing designer's portfolio would include a mitten that accommodates a lone forefinger in firing position.*

In an Army gone increasingly high-tech, the modern military uniform is less an outfit than a system. It's a holder of gizmos and gear and the ammunition and batteries that go in that gear. Back before hulking body armor and gear-festooned vests supplied the intimidating profile, the clothing itself was sometimes recruited for the task. High hats and epaulettes made officers appear taller and more broad-shouldered. And the boots. *The boots.* Dashing knee-high leather boots protected the pant legs, yes, but surely they also boosted morale. Uniforms created not just uniformity but brio and self-confidence. They were crisp, flattering, finished with piping

---

* The Cold Weather Trigger Finger Mitten Insert, a component of the Extreme Cold Weather Mitten Set. As opposed to the Intermediate Cold Wet Glove or the almost fashionable-sounding Summer Flyer's Glove.

and grosgrain and tassels. They were, to quote Annette LaFleur, "very couture."

The current combat uniform, with its sensible emphasis on hot-weather comfort, is worn loose and untucked. It doesn't say "ready to kill" so much as it says "ready for bed." Still, clothing remains an important Army morale issue. The ACU used to be unisex, but women complained. The shoulders and waist were too wide for many women, or the hips too narrow. The knee patches tended to hit at the shins. Women hated it. They hated it enough that the Army commissioned a female uniform.

"But you can't call it that," Accetta says. "Because some of the guys are wearing them." It's called the Army Combat Uniform–Alternate, a uniform "for smaller stature Soldiers."

Every now and again, military fashion has evolved not out of practicality or research or matters of morale but simply out of the sartorial inclinations of one high-ranking individual. British history holds a General Cardigan and a General Raglan, and I like to picture them in their tented quarters, sketching outerwear by lantern light. Most recently, an Army chief of staff decided that black wool berets would be the headpiece of the Army Combat Uniform, not so much because wool is flame-resistant and moisture-wicking but because he dug the look. He dug it despite having to wrangle an exception to the Berry Amendment. He dug it despite his troops' near-universal preference for visored cloth caps. And despite their having every good reason to prefer caps, because not only do caps shade the eyes but they're cooler than berets, and lighter and less bulky in a pants leg pocket. (It took ten years, but the Army has finally got its caps back on.)

The most talked about—around Natick—example of sartorial rank-pulling was the Universal Camouflage Pattern used on the

Army Combat Uniform beginning in 2005. The idea had been to develop a single camo pattern that would provide concealment for troops in desert, urban, and woodsy settings. The Natick Camouflage Evaluation Facility came up with thirteen pattern and color combinations, duly sent overseas for field tests and feedback. Before the data was in and the study completed, a highly placed general went ahead and picked a pattern. It was not even one of the ones being tested. The new camouflage performed so poorly in Afghanistan that in 2009, the Army spent $3.4 million developing a new and safer pattern for troops deployed there.

Camouflage is interesting from a fashion perspective. As a rule, the military starts—rather than follows—fashion trends in the civilian sphere. Every now and then, they start them *and then* follow them. Midway through the previous century, Army camouflage prints began showing up in mainstream fashion. It began with clothing and took off from there. As I write this, you can get on the internet and order camouflage wedding bands, dog sweaters, onesies, condoms, flip-flops, hard hats, and football cleats. Camo print became so popular that eventually Navy personnel began clamoring for it. To the embarrassment of many, the current Navy working uniform is a blue camouflage print. Unsure whether perhaps I was missing the point, I asked a Navy commander about the rationale. He looked down at his trousers and sighed. "That's so no one can see you if you fall overboard."

No military fashion foolery can compete with the saga of the orange-red underwear. Around the turn of the last century there was, in the phrasing of a paper in the June 1897 issue of *Medical Bulletin*, "a very prevalent idea that red underwear possesses some occult medical value." Baseless as it was, the notion made its way to the office of the US surgeon general. A Lieutenant Colonel William

Wood reported that British Army officers in India had found some relief from the intense tropical sun by lining their hats with red cloth. A study was commissioned, using troops stationed in the Philippines as subjects. Though a small number of caps were prepared, the Americans seized on the idea of red underwear—perhaps as a sort of secret asset, a hidden psychological edge, like fabulous lingerie or lifts in one's shoes. Five thousand pairs of orange-red drawers and undershirts were shipped from a War Department depot in Philadelphia, along with a like number of white drawers, as controls. Mental and bodily vigor to be monitored for one year. One thousand men conscripted as subjects.

The garments arrived on the first of December 1908. Whereupon the setbacks began. Four-fifths of the underwear was too tight for all but the smallest men. Perhaps, having seen the destination of the shipment, the clothiers had misunderstood and fashioned the items for the slighter, trimmer bottoms of the Filipino male. Perhaps they were cutting corners. Who can say. Six hundred men were dropped from the study. Worse, the underpants had been sewn from heavy dungaree cotton, causing the men to perspire in places where no one wants to perspire, surely counteracting any mystical cooling properties that the color might have conferred. The sweating proved doubly vexing, as the dye ran badly. Much joshing and ridicule—"bantering by their companions"—ensued. Over a month of washings, the red underwear progressed to yellow and onward to a "dirty cream color." The cap liners, though less frequently laundered, ran in the rain or when the men sweated, causing red blotches and rivulets running down their faces and yet more bantering.

At the end of a year, the men were interviewed about their experience with the special underwear. Only 16 of 400 had anything positive to say. The red underwear was hotter and scratchier. It "outraged the sensibilities of the wearer." In addition to prickly heat and rashes and a higher rate of heat exhaustion, it was blamed for headaches, dizziness, fever, blurred vision, boils, and colic. "An Experiment with Orange-Red Underwear" was read aloud at a biennial meeting of the Far Eastern Association of Tropical Medicine, no doubt jollifying the usual assortment of malaria and foot fungus reports.

While we are on the topic of hot, uncomfortable military intimates, a word about Kevlar underwear. A version of which was successfully marketed to the British Army: Blast Boxers*—protection against *life-altering* injuries.

"There was some controversy over this," Accetta says.

LaFleur's large, beautiful eyes widen. "She's recording you."

"Which I started."

Fox News had called, asking why the Brits had bombproof underpants and our guys didn't. "No such thing," Accetta told the reporter. He said they were working on something else, and then he got off the call and dialed the Natick ballistic protection guys. What in fact were they working on? Silk underwear, said the ballistics guys. No, *really*, they said. Unlike other naturally breathable textiles—cotton, say— silk doesn't fragment and riddle a wound with bits of fiber that can

---

* Referred to casually as the "combat diaper" or "blast diaper." But not "codpiece." Possibly because codpieces likely had nothing to do with genital protection—or fashion, for that matter, or cod. C. S. Reed, writing in the Occasional Medical History Series of the *Internal Medicine Journal*, speculates that codpieces were worn to cover syphilitic buboes (swollen, inflamed lymph nodes) and the "bulky woolen wads" used to absorb the "foul and large volumes of mixed pus and blood . . . discharged from the genital organs." It's all speculative, because the fabric codpieces, the pus, and the woolens have all disintegrated in the intervening centuries. We do have Henry VIII's suit of armor, which features a codpiece like the nose of a Cessna, but historians now say there's no evidence the king had syphilis. The only thing I can say for sure is that Bubo (in Kuwait City) is an even less appealing name for an eating establishment than Bursa (San Francisco).

seed infection. It's surprisingly strong. Spider silk, specifically, has a better strength-to-density ratio than steel. (Natick at one time had a "spider room" in the basement of Building 4 and a team of scientists working to parse the unique protein structure of spider webbing in order to synthesize it.)

Despite its strength, silk doesn't inspire confidence. Kevlar does that. People don't realize that a khaki-weight Kevlar (or its newer cousins Spectra and Dyneema) doesn't stop bits of metal packed into improvised explosives. For that you'd need fifteen to forty plies—far too heavy for underwear. What Blast Boxers stop is dirt and sand launched by buried bombs. That is important, because dirt carries fungi and bacteria that can cause deep-lying, hard-to-conquer infections in a wound. Blast Boxers are excellent. But they are most assuredly not, as the advertising material implied, bombproof. "Honestly," Accetta says, "if the insurgents can make a bomb big enough to blow up a seventy-ton M1 tank, they can certainly make a bomb that's going to blow up your underwear."

2

Boom Box

Automotive safety for people
who drive on bombs

A S SOMETIMES HAPPENS IN rural America, someone has shot up a road
sign. The sign—a right-turn arrow on a yellow background—
stands on a paved lane along the edge of Chesapeake Bay. Given
the gape of the hole and the fact that the road traverses Aberdeen
Proving Ground, there's a good chance it wasn't made by a bullet.
A proving ground is a spread of high-security acreage set aside for
testing weapons and the vehicles meant to withstand them. In the
words of the next sign up the road: Extreme Noise Area.

I'm headed for Aberdeen's Building 336, where combat vehicles
come to be up-armored—as the military likes to up-say it—against
the latest threats. Mark Roman, my host this morning, oversees the
Stryker "family" of armored combat vehicles. He'll be using them for
an impromptu tutorial in personnel vulnerability: the art and science of
keeping people safe in a vehicle that other people are trying to blow up.

My extremely uneducated guess is that some sort of shaped charge
hit that sign. A shaped charge is an explosive double whammy used
for breaching the hulls of vehicles and harming the people inside

them. The first blast propels the murderous package to its target. On arrival, the impact detonates a wad of explosives packed inside. The blast slams a metal disc positioned in front of the explosives. Combined with the weapon's contour, the energy of the blast shapes the metal into an extremely fast, close-range projectile that can punch through the hull of an armored vehicle with little trouble. RPGs (rocket-propelled grenades) are the ones most people have heard of, though there are ever bigger, deadlier iterations. Defense Industries of Iran is said to have one that can push through 14 inches of steel. Shooting a traffic sign with a shaped charge is like using a leather punch on a Kleenex.

By and large, an army shows up to a war with the gear it has on hand from the last one. The Marines arrived in Iraq with Humvees. "Some of the older ones had *canvas doors*," says Mark, who was one of those Marines. His hair has since gone silver, but he's retained the ready, let's-do-this physicality that the Marine Corps seems to impart. When I asked a question about a new blast-deflecting chassis, he grabbed some wheeled mechanics' boards and we rolled beneath a Stryker and finished the conversation on our backs.

Early on in Iraq, the Army tried plating vehicles with MEXAS armor panels, which work well against heavy machine-gun fire. "We were like, *crap*," Mark recalls. "*This does not stop an RPG.*" You might as well have armored your vehicle with right-turn signs. Another thought was to add tiles of reactive armor, a sort of exploding Pop-Tart affair. When an RPG hits it, the filling explodes. This outward-directed blast serves to negate the blast of the RPG—and any passing pedestrian. Given that much of the fighting during the first Iraq conflict took place in urban areas—and was ostensibly an effort to "win hearts and minds" among the populace—reactive armor would have been a poor choice.

Besides, something cheaper and simpler had been found to work. Mark rolls out from under and leads the way to another Stryker, this one in a hoopskirt of heavy-duty steel grating called slat armor. The nose of an incoming RPG gets pinched between two slats, which duds it. It's like squeezing your nose to stop a sneeze: It either prevents the explosion from happening or blocks the expulsion of nasty stuff. Either way, it proved effective. Strykers would lumber back to base like up-armored hedgehogs, bristling with RPGs. Slat armor worked so well that Iraqi insurgents largely gave up on RPGs.

And switched to making bombs. Early on in the Iraq war, improvised explosive devices were hidden on the sides of roads. Since these IED blasts hit vehicles broadside, the Army responded by flanking them with armor plates and replacing windows with "Pope glass"—two-inch thick transparent armor of the type that keeps His Holiness whole on his own tours of duty. Better, but it left the machine-gun turret exposed. Platoons tried piling sandbags up there, but they'd burst apart and literally sandblast the gunner. More ballistic shielding was added.

And thus more weight. All the added armor had Humvee engines screaming and straining on the uphills, and brakes burning out on the downs. Safety modifications on the Strykers added 10,000-plus pounds—far more than the vehicle was built to handle. You can beef up the suspension and tires, replace the engine—all of which was done—but you've still got problems. Past a certain tonnage, an armored vehicle begins to Godzilla the landscape. It breaks up asphalt, collapses levees. Exceeds the cargo capacity of the planes that deliver it. For every piece of armor and reinforcement, people like Mark would be called on to ditch something of similar weight. And the Stryker was never a lushly appointed vehicle. There is no onboard toilet. (There are empty Gatorade bottles.) The early ones didn't even have air-conditioning. I tell Mark I'm glad to see some cup holders were left in place. I recognize the brief, polite silence that follows. It's Mark Roman rendered mute by the fullness of my ignorance. They're rifle holders.

Fast-forward to Afghanistan: land of the hundred-pound IED. To get around the up-armoring, insurgents came at vehicles from below, burying the explosives in the middle of the road rather than on the sides of it. As on most trucks, the chassis on US combat vehicles at that time were flat. Where newer generations of vehicles have V-shaped or double-V-shaped chassis to deflect the energy unleashed in a blast, the flat ones took it head-on. And because the seats were bolted to the passenger compartment floor, the energy would transmit directly to passengers' feet, spines, and pelvises. Smacked them bad.

Newer vehicles have higher clearance. The force of a blast diminishes rapidly as it radiates outward. The energy at one or two feet is still so condensed that it can act like a solid projectile and break through vehicle floors. Once the integrity of the hull is breached, any loose piece of vehicle or gear becomes a projectile. Soldiers and Marines would pile sandbags on the floors of Humvees for the same reason aviators used to sit on their body armor instead of wearing it. Because death came up from below.

The underbody blast scenario was dire enough that US Central Command rolled out the procedural big guns: They issued a JUON (say, *joo-on*): a Joint Urgent Operational Need Statement. The statement surely ran longer than fifteen words, but the gist was this: Get us some combat vehicles that can drive over bombs and keep everyone inside alive. Nine vendors submitted prototypes for what would come to be known as MRAPs (say, *em-wraps*): mine-resistant, ambush-protected. But without fielding them first, how do you know which one is safest—and precisely how safe it is? You hire a "personnel vulnerability analyst."

The Army Research Laboratory snapped up Nicole Brockhoff, premed at Johns Hopkins, with a graduate degree in biodefense. The youngest person to win the Secretary of Defense Meritorious Civilian Service Award. Bench presses 190. She's come down from her office in the Pentagon to attend to some things, and agreed to make me one of them. Whenever Mark takes over the explaining, Brockhoff drops back and takes out her phone. She does not seem rude, just grindingly busy and determined to stay on top of her day. I see her come and go in my peripheral vision, pacing, answering email. She gives the impression of someone for whom idleness is almost physically unbearable. She is gorgeous, articulate, fast-moving, powerful. Lesser humans left blinking in her wake.

Brockhoff offers to show me another anti-IED modification: the energy attenuating seat. We climb inside the passenger compartment of a Stryker infantry carrier, which does not have a door but rather a drop-down ramp, like a circus boxcar. The first good thing about these new seats is that they are no longer bolted to the floor. Second, they ride on special shock-absorbing pistons. What's special are the collapsible, replaceable metal inserts that slow the seat's downstroke and keep it from bottoming out. The catch is that in order for passengers to protect their feet and lower legs, they need to keep them off the floor. The footrests on the base of each seat are for the person sitting directly across. Meaning that one soldier has to straddle the other's knees for hours at a time. Mark, who has joined us, adds that having the knees up like that tends to make the butt go numb. "Like when you're reading on the toilet too long. And you get toilet palsy."

The last two words hover, finding nowhere to touch down. "Man thing," Brockhoff decides.

On a long drive, fighters' feet surely stray from the safety of the footrests. But their commanders likely know which parts of the route are riskiest and can give a heads-up.

Speaking of heads and up, I ask about airbags on the ceilings, to prevent brain injury. Unfortunately, automotive airbags don't respond quickly enough to get the jump on a blast. Early on in her tenure at the Pentagon, Brockhoff found herself talking to a general about the challenges of high-speed energy mitigation. He suggested she talk to NASCAR.

"I said, 'With all due respect, General. . . .'" The bottom of a personnel carrier is traveling many, many times faster than a NASCAR race car. And unleashing a force of many times greater magnitude. Besides, NASCAR's approach won't work for combat vehicles. Race car drivers are packed in their seats like mail-order stemware. Heads are braced and supported, so necks don't break and brains don't ricochet against skulls. Danica Patrick can't even look out the driver-side window and wink at the pit crew. That's no good for combat vehicles. Drivers and gunners need to be scanning in all directions, looking out for suspicious elements: piles of trash or dead goats that might be hiding bombs, people holding cell phones that might be wireless detonators, children with their fingers in their ears.

At the same time that the Army was working to make existing vehicles safer, they were scrambling to evaluate the new MRAPs. When Brockhoff arrived, her colleagues were using the crash test dummy that the auto industry uses: the Hybrid III. First, because that's what there is. And second, because it makes some sense. Both a car crash and an underbody blast cause blunt force trauma: the sorts of injuries you get from slamming into pieces of a vehicle's interior. (As opposed to injuries caused by a blast pressure wave passing through you—rupturing organs and eardrums and the like—which a vehicle largely protects against.)

Here's the problem: automotive crash test dummies were designed for measuring force mainly along two axes—front to back (for head-on impact), and side to side (for "T-bone" crashes). With a blast coming up from below, the axis of impact runs vertically through the body: heels to head. "This doesn't," Brockhoff told her colleagues gently, "seem like it's going to be sufficient moving forward." To make the point, a Hybrid III was filmed alongside a cadaver in a controlled blast. It is clear, from the slow-motion footage, that this dummy wasn't built for this. It's like watching an elderly, arthritic man try to follow along in a Zumba class. Compared with the flailing arms of the cadaver, the dummy's barely move. When the real head comes down, the dummy's is coming up. Its thighs rise a third as high off the seat as the cadaver's, and its ankles barely flex.

The Hybrid III captures the basic pattern of injury—feet, lower legs, spine—but it doesn't provide the level of detail Brockhoff's team needed. "We were missing a lot of nuance about the severity of the injury. We needed to know, at what point do you go from a treatable injury that's recoverable to something life-altering and incapacitating and potentially fatal? We need to be able to make those distinctions when we're testing these trucks. And we can't right now."

So the Army is building a dummy of its own. WIAMan—the Warrior Injury Assessment Manikin—will be specifically tailored for underbody explosions. The project employs about a hundred people (most of whom, as far as I've been able to determine, have never watched *Jackass* and thus had no knowledge of the dwarf cast member Wee Man).

WIAMan is starting the way the automotive crash test dummy people started: with cadavers and bioengineers and controlled blasts of varying magnitude, followed by autopsies to document the injuries. Before they could start any of that, they had to build a blast rig, something robust enough to withstand an explosion directly below it. The tower, as it is conversationally known, stands in a meadow near what the mapmakers call Bear Point and the Aberdeen Explosive Effects Branch calls Experimental Facility 13. I am headed over to EF13 after lunch. The cadavers are there already, sitting in seats on the tower platform. They arrived a day ago from bioengineering labs at three universities. Some made the trip in a modified horse trailer, disappointing the children in passing cars craning their necks for a glimpse of tail or rump.

EF13 IS lovely this time of year. A late October sun softens the chill and highlights the white butterflies that flit around the bioengineers as they work. The clearing is edged by oaks, changing their outfits before dropping them to the floor. The cadavers too, wear fall colors, one in an orange Lycra bodysuit* and one in yellow. For now, they sit slumped in their seats, chins on their chests, like dozing subway commuters.† Because the setup takes two days, the dead men spent the night in the meadow. A portable weather shelter was erected to protect the electronics, and a pair of guards took

---

* I emailed Vandue Corp, one of the companies that sell full-body Lycra suits, to see if they were aware of having tapped the cadaver apparel market. The customer care person replied that they were not. Though word had reached them that their product had caught on with bank robbers, as the face is covered but allows the wearer (if living) to see out. Presumably the felon, unlike the Halloween revelers and sports fans who more routinely don Lycra suits, wore some clothing over his. Though I hope not. And I further hope he selected the Sock Monkey pattern.

† Sleeping subway riders, conversely, look exactly like dead men—a fact born out by the regular appearance of news items about commuters who quietly die and then sit, slumped and unnoticed, through several round-trip circuits of the route. As a passenger quoted in "Corpse Rode the No. 1 Train for Hours" attests, "He just looked like he was asleep."

turns watching from a truck parked nearby. Bear Point may not have bears anymore, but it does have coyotes, and neither death nor Lycra dampens a coyote's enthusiasm for meat.

Under the platform is a small plot of simulated Middle East: engineered soil that has been heated and moistened as per protocol. Consistency and repeatability being key elements of the work. At around 2:30 p.m., a pickup truck will arrive with a few pounds of the explosive C-4, which everyone here has been referring to as "the threat." Around 2:45, the bioengineers and investigators and hangers-on like me will be escorted to a nearby bunker while the threat is buried in the special dirt and a detonating wire is attached. Then the wood staircase to the tower platform will be pulled away (so the carpenters don't have to keep rebuilding it), and an alarm will sound three times. After which the threat becomes the event. The Tower, the Threat, the Event. It's like a tarot deck out here.

It's just past noon now. The cadavers are having their connectivity rechecked after the long drive in. Data will be gathered from sensors mounted on their bones and then transmitted along wires laid down along the insides of their limbs and spines—a sort of man-made nervous system. As with the real deal, the nerves lead to a brain, in this case the WIAMan Data Acquisition System. A bundle of wires exit at the back of each specimen's neck and feed into the system.

After the blast, the cadavers will be autopsied and the injuries documented. This is the information that will allow vehicle evaluators to interpret the g-forces and strains and accelerations that WIAMan's sensors will register. Because of the cadavers' contributions, WIAMan will be able to predict what kind and what degree of injury these different magnitudes of force would be likely to cause in an actual explosion. WIAMan won't be done until 2021, but in the meantime, the cadaver injury data can be used to create a transfer function, a sort of auto-translate program for the Hybrid III.

By now the cadavers have been coaxed into a straight-backed dinner-table posture, some duct tape keeping them from slumping. (In coming months, data will be gathered for more realistic positions—legs stretched out in front or angled back under the seat.) A bioengineer holds one of the heads in his hands, like a man in a movie preparing to kiss his co-star. Another strings thin wires to hold the head in that eyes-right position, though not so firmly that it interferes with its movements, which will be captured on video cameras set up in bunkers on all four sides. There's a protocol for everything: the angle of the cadavers' knees, the position of their hands on their thighs, the newtons of force with which their boots are laced.

The bucolic calm of the setting belies the pressure everyone's under to get the bodies prepped on schedule. A butterfly lands, unnoticed, on a bioengineer's shoulder. Jays converse, or seem to, with the scratchy calls of duct tape being pulled from the roll. The hover and fuss of the scientists exaggerates the abiding stillness of the bodies. They're like anchormen sitting for their makeup. How nice for them to be outdoors on this fine, crisp autumn day, I find myself thinking. How nice to be in the company of people who appreciate what they've agreed to do, this strange job that only they, as dead people, are qualified to do. To feel no pain, to accept broken bones without care or consequence, is a kind of superpower. The form-fitting Lycra costumes, it occurs to me, are utterly appropriate.

Not everyone feels the way I do. In 2007, someone at the Pentagon complained to the Secretary of the Army about a preliminary WIAMan test. "I'll never forget," says Randy Coates, WIAMan's project director until his retirement in 2015. "It was a Wednesday evening, about seven o'clock. I got a call from a colonel over at

Aberdeen, where we were going to run the test. He says, 'The Secretary of the Army has shut down the test.' We had three cadavers
and a team of people who'd been working on them around the clock
for days." As Brockhoff recalls it, "Someone felt their personal beliefs
had been affronted." Her boss went to the Secretary and tried to
explain: You can't build a human surrogate without understanding
how the human responds. And then he got mad. To shut down the
project at the last moment like that would be not only an extrava-
gant waste of money but a waste of the donors' bodies. Sometime
on Friday, the last possible day before decomposition would have
invalidated the results, the test was cleared to go forward—surely
the first cadaveric research venture with multiple two- and three-
star generals in attendance.

Jason Tice, who oversees WIAMan live-fire testing, pointed out
that the sudden, intense scrutiny may have had a silver lining. "It's
been informing leadership about the risks they're subjecting soldiers
to." In other words, my words, maybe they'll worry a little less about
the dead and a little more about the living.

The downside to the Pentagonal hullabaloo is a newly bloated
approval process. The protocol for research involving cadavers has
to be approved by the head of the Army Research Laboratory and
by ARL's overseeing organization, the Research, Development and
Engineering Command. From there it goes to the commanding general of the Army Medical Research and Materiel Command, which
in turn passes it on to the Surgeon General of the Army, who sends
it to Congress. Who have two weeks to respond. And if no one along
the way takes issue, then and only then can the work begin. The
whole process can take as much as six months.

The other fallout is a newly drafted "sensitive use" policy.
Potential body donors are required to have given specific consent

for research or testing that may involve, as the document lays it out, "impacts, blasts, ballistics testing, crash testing and other destructive forces."

Who would sign such a thing? Plenty of people. Sometimes, Coates says, it's people who like the idea of doing something to help keep military personnel safe. It's a way of serving your country without actually enlisting. I can imagine there are people who, while drawn to the nobility of risking life and limb for a greater cause, would prefer to do so while already dead. Mostly, I'm guessing, it's the same sorts of people who donate their remains for any other worthy endeavor that relies on the contributions of the insensate. If you're fine with a medical student dissecting every inch of you to learn anatomy, or with a surgeon practicing a new procedure or trying out a new device on you, then you are probably fine riding the blast rig. *I won't be needing it,* is the typical donor attitude toward his or her remains. *Do what you have to do to make good from it.*

I**N WORLD** War II they called it deck-slap. Explosions from underwater mines and torpedoes would propel a ship's decks upward, smashing sailors' heel bones. Like "combat fatigue" for post-traumatic stress disorder, it was a cavalier toss-off of a name for what would often turn out to be a life-altering condition. The calcaneus (the heel) is tough to break, tougher still to repair. By one early paper's count, eighty-four different approaches had been tried and discussed in medical journals. Dressings of lint and cottage cheese. "Benign neglect." "Mallet strikes to break up fracture fragments" followed by "manual molding" to recreate a heel-like shape. Few statistics from the era exist, but one paper cites an amputation rate of 25 percent.

Underbody blasts have brought heels back to the attention of military surgeons. The mallets and lint have been replaced with surgery and pins, but the amputation rate for deck-slap injuries is higher than ever—45 percent, in one recent review of forty cases. Part of the problem has to do with fat, not bone. The calcaneal fat pad keeps the bone from abrading the skin on the underside of the heel. It's an extremely dense, fibrous fat found nowhere else in the body. (There's enough squish there to merit the cobbler's term "breast of the heel.") Fat pads are frequently damaged in underbody blasts, sometimes badly enough that they have to be removed. Without the padding, the pain of walking is acute. When vitamin A poisoning caused the soles of Antarctic explorer Douglas Mawson's feet to slough off, he stuffed them in the bottom of his boots like Dr. Scholl's cushioning insoles. It was the only way he could go on.

Can't something be put in to replace a damaged fat pad? I spoke to orthopedic surgeon Kyle Potter, who works with these patients at Walter Reed National Military Medical Center. "You mean like a small silicone breast implant?" I wasn't actually thinking that, but sure.

"No." Potter pointed out that breast implants aren't designed to stand up to the forces of heel strike. Walking pounds the calcaneus with 200 percent of a person's body weight; running, as much as 400 percent. Rupture and leakage would likely be issues. At best, Potter said, it would feel very strange. It would feel like someone stuck a breast implant in your shoe. And who, other than Douglas Mawson, would want that?

In half an hour, some deck-slap will be broadcast live on the video monitors in the bunker. We're all over there now, while the explosives team readies the bomb. There's not much else in here. Some microwave ovens for warming engineered soil ("DIRT ONLY," they are labeled). An earplug dispenser by the door. The plugs are pastel foam, shot through with sparkles. It seems like a lot of manufacturing bother just to be able to call your product Spark Plugs. A wall clock shows the wrong time. "No one can figure out the admin system for the clocks," a man explains. "We can't spring forward and fall back."

We stand and stare at the video feed. A slight breeze moves the trees beyond the tower. Someone with a working timepiece begins a countdown. The explosion sounds muffled, less by earplugs than by distance. We're a half-mile away. The cadavers appear to be thrown by the blast, but not in an action-movie way. More of a took-a-speed-bump-too-fast way. As with an automotive "crash test," the language is more disturbing than the actual event. The cadavers in an under-body blast test are blown up, as in *upward*, not *apart*.

The event is filmed at 10,000 frames per second. Playing back the footage at 15 or 30 frames per second allows the researchers to step inside the half-second lifespan of the event. Now we can see what in real time we could not. First the boots flatten, their sides bulging noticeably. An index finger rises from where it was resting on a thigh, as though the cadaver were about to make a point. The lower legs extend and rise. The head comes down and the arms shoot out in the manner of a hurdler mid-leap. Coates reverses the footage and directs me to watch the spine. As the energy of the blast moves to the seat pan, the dead man's pelvis rises, shortening his torso and expanding his paunch. Underbody blast can compress a seated soldier's spine by as much as two inches. Back pain and injuries, no surprise here, are common.

Played at this speed (and in this outfit), it's modern dance. There's grace and beauty to the limbs' extensions, nothing brutish or violent. In real time, though, the forces that move the limbs pass too quickly for the tissue to accommodate. Muscles strain, ligaments tear, bones may break. Imagine pulling apart a wad of Silly Putty. Pull slowly, and it will stretch across the room. Yank it fast and it snaps in two. Likewise, different types of body tissue have different strain rates. For the forces of any given blast, one type may stretch, say, a fifth of its length without tearing, while another may manage just 5 percent. WIAMan will be calibrated to reflect these differences and predict the consequences.

The long-term quality of a soldier or Marine's life is a relatively new consideration. In the past, military decision makers have concerned themselves more with go/no-go: Do the injuries keep a soldier from completing her mission? Have we lost another pawn in the game? WIAMan will answer that question, but it will answer others, too. Is this soldier likely to have back pain for the rest of his life? Will he limp? Will his heel hurt so much that he'd rather lose the foot? The answers may or may not affect the decisions that are made, but at least they'll be part of the equation for those inclined to do the math.

BACK IN Building 336, I ask my hosts if it would be okay to try driving a Stryker. It would not. Like an obliging parent, Mark allows me to sit in the driver's seat and turn the steering wheel back and forth for a minute before he reparks the thing. Brockhoff, pacing at the edge of the parking lot, has found some sort of plastic packing material. She darts over to the Stryker and stuffs it in behind the backmost tire. What follows is a noise you will find nowhere in the publications of military hearing professionals: a 40,000-pound Stryker backing up over an armload of wadded-up bubble wrap.

3

Fighting by Ear

The conundrum of military noise

THE UNITED STATES MARINE Corps buys a lot of earplugs. You find them all around Camp Pendleton: under the bleachers at the firing range, in the bottoms of washing machines. They are effective, and cheap as bullets* (which also turn up in the washing machines). For decades, earplugs and other passive hearing protection have been the main ammunition of military hearing conservation programs. There are those who would like this to change, who believe that the cost can be a great deal higher. That an earplug can be as lethal as a bullet.

Most earplugs reduce noise by 30-some decibels. This is helpful with a steady, grinding background din—a Bradley Fighting Vehicle

---

* And, though you didn't ask for it, here's one more similarity between bullets and earplugs: Both have been used by physicians to protect their ears from screams. The *Army Medical Department Journal* states that the real reason soldiers in the pre-anesthesia era were given a bullet to bite was not to help them endure the pain but to quiet their screams. And from a paper called "The History and Development of the E-A-R Foam Earplug" we learn that emergency room docs use foamies "to block the screams of children during difficult procedures." This was part of a section on "unusual applications," none of which were especially unusual. I may have had unreasonable expectations for the history of the foam earplug.

clattering over asphalt (130 decibels), or the thrum of a Black Hawk helicopter (106 decibels). Thirty decibels is more significant than it sounds. Every 3-decibel increase in a loud noise cuts in half the amount of time one can be exposed without risking hearing damage. An unprotected human ear can spend eight hours a day exposed to 85 decibels (freeway noise, crowded restaurant) without incurring a hearing loss. At 115 decibels (chainsaw, mosh pit), safe exposure time falls to half a minute. The 187-decibel boom of an AT4 anti-tank weapon lasts a second, but even that ultrabrief exposure would, to an unprotected ear, mean a permanent downtick in hearing.

Earplugs are less helpful when the sounds they're dampening include a human voice yelling to get down, say, or the charging handle of an opponent's rifle. A soldier with an average hearing loss of 30 decibels may need a waiver to go back out and do his job; depending on what that job is, he may be a danger to himself and his unit. "What are we doing when we give them a pair of foam earplugs?" says Eric Fallon, who runs a training simulation for military audiologists a few times a year at Camp Pendleton. "We're degrading their hearing to the point where, if this were a natural hearing loss, we'd be questioning whether they're still deployable. If that's not insanity, I don't know what is."

Fallon is lecturing in a classroom at the moment, but after lunch the audiologists in attendance will experience some live-fire simulated combat. Working with the Department of Defense Hearing Center of Excellence, Fallon contracted a company, ArmorCorps, who in turn brought in a team of Marine Corps Special Operations forces, and together they've set up a half day of warfare scenarios. The aim being to provide the hearing professionals with a firsthand feel for the frustrations and dangers of the current approach, with the hope that they'll become advocates for something better.

Fallon turns the class over to ArmorCorps' Craig Blasingame, a former Marine with a wide superhero jaw and muscles so big that when he walks in front of the slide projector, entire images can be viewed on his forearm. Though it's ten in the morning, Craig has a five o'clock shadow.

"We're going to put you in an environment today and show you what it's like to try to maintain a degree of auditory situational awareness while you're wearing passive hearing protection." Craig speaks like a bullhorn. He says this is because he lost some hearing as a Marine, but I think it's because there's so much strength coursing around in there that everything—whiskers, voice, the pectorals under his polo shirt—wants to burst forth in a powerful way.

Craig and Aaron Iwanciw, also a former Marine and now CEO of ArmorCorps, will shortly be taking the audiologists (and myself) outside for a listening exercise. Because we're going to the firing range afterward, this is one of the rare listening exercises that will be done in body armor and a combat helmet. Aaron is smaller and quieter than Craig, and smells pleasantly of shampoo. He helped me put on my gear. ("I can stick my lip balm and tape recorder in these little skinny vest pockets." "Those are for ammo.")

Outside the building, Craig arranges us in a patrol formation. In a war zone, just walking down the road has a strategy to it. The "killing radius" of a fragmentation grenade is 15 feet. If troops walked along in a clump like tourists, a single grenade could kill the lot of them. Thus they keep fifteen to forty-five feet between each other on patrols. The more spread out they are, though, the harder it is to hear one another, especially with hearing protection in place.

Aaron is the point man, while Craig takes up the rear. We're wearing ear cuffs similar to the ones people wear when they operate power tools. Craig sounds like someone shrank him down and put him inside a mason jar. I believe I hear him tell us to "step off right." I interpret this as meaning we're all supposed to clear off onto the right side of the road, so I start crossing over. "You're on the left," barks a teammate. He sounds sure, so I turn back and narrowly avoid being hit by a black Suburban creeping along seemingly out of nowhere. When a two-ton SUV driving on gravel can sneak up on you, that's not good.

Fallon says lapses in communication were the norm when he was in the infantry. "Do you know how much time I spent saying, 'I have no idea what's going on'? We'd get an order to halt, spread out, take up a more hidden position. My buddy next to me would be going, 'What's going on?' And I'm going, 'Shit, I don't know what's going on.'" And you couldn't yell, 'Hey, what's going on?' because the enemy would hear you and know where you were.

Aaron is leading the next exercise, a live-ammo "tactical scenario" out in the wilderness beyond the firing range. Before we head over, he has us push a button on the ear cuffs we're wearing. This is the point in my notes where it says, "bionic!" As a kid, I used to watch a TV show called *The Bionic Woman*. Like her male counterpart of an earlier season, she'd been rebuilt by the military with experimental superprosthetics following some variety of hideous maiming. It was the least they could do. One of the implants was for her ear. She'd cock her head, and suddenly we'd be eavesdropping on a pair of underworld kingpins in a Buick Riviera across the street. I have her hearing now. Aaron is fifteen feet away, talking to Craig, but he sounds so close I should be smelling shampoo.

The name for what we've got on is TCAPS (say, *tea-caps*), Tactical Communication and Protective System. Incoming noises are analyzed; the quiet ones are amplified and the loud ones reproduced more quietly. (The system also incorporates radio communications, or "comms.") So far it's mainly Special Operations forces who are using TCAPS. Why? Money, of course, but also the fact that it comes out of the radio budget, and the majority of foot soldiers don't carry radios. Plus some skepticism among leadership. "Senior NCOs," says Fallon, referring to noncommissioned officers, "will flat-out tell you, 'Don't give me more shit that's supposed to be the next high-tech wonder that's going to break or the batteries are going to go dead and I've got to carry it.'"

The shit we have on happens to be made by Fallon's employer, 3M. "I hope you don't think that that's what this is about," he said to me at one point. I don't, entirely, no. Fallon is an evangelist for the product category, not the brand. 3M also supplies earplugs to the military, so either way, they've got a tasty piece of defense budget pie.

The hearing professionals and myself are joined for the tactical exercise by twelve men from a Marine Corps Special Operations unit, the name of which I've been asked to omit. Aaron briefs the lot of us.

"You are a Special Operations team heading into a village in Afghanistan," he begins. "The mission is to make a liaison with the village elders. Engage the elders, ask about Taliban activity in the area. Ask them about their quality of life. What their problems are." Perhaps fit them for hearing aids. "In support of the operation, we have a Predator drone in the overhead, and quick access to an assault weapons team: Cobras or Hueys. If things go kinetic we can call them up for supporting fire." *Going kinetic* is military shorthand for *people are firing guns at you*. In this case, they're imaginary people, but the Spec Ops guys will be shooting back anyway, because this is an exercise about communicating in the chaos and clamor of combat.

We're instructed to turn our radios to channel 7 and line up behind one of the Special Ops guys, two of us per guy, as close as possible without hitting his boot heels. "If he runs, you run," says Aaron. "If he takes a knee, you take a knee." Myself and a middle-aged audiologist with braids poking down from her helmet get behind a short man who is hard to describe because all distinguishing features except his nose are obscured by gear of some kind. He introduces himself and says hi.

"Hi, I'm Mary," says the audiologist.

Me too, I say. "I'm also Mary."

"Well," says our Special Ops guy, clearly unaccustomed to so much Mary. "That does make it easy for me."

We set off into the scrub. Camp Pendleton is two hundred square miles, with seventeen miles of California coastline, much of it left wild for practice invasions and amphibious assaults. It's like a national park reserved for the U.S. Marine Corps and a lot of twitchy wildlife. (The grunts are forbidden to shoot the animals, but I'm guessing it happens. I'm guessing this because I recently visited the Camp Pendleton paintball range and asked to be shot to see what it feels like. Fifteen Marines volunteered. The one who did the deed—from 70 feet, hitting me precisely where he wanted to—can be heard in the background of a researcher's video going, "That was *very* satisfying.")*

As we make our way across the terrain, a multiparty conversation unfolds in my ear cuffs. One man is talking with the drone operator, and someone else is communicating with the Cobra pilot and the attack controller. Everyone, including the President of the United States, if he wished to, can switch their comms to channel 7 and listen in. (When Navy SEALs stormed Osama bin Laden's

---

* "It's almost like he knows you," said the researcher.

compound, they were wearing TCAPS, and President Obama and Secretary of State Hillary Clinton were listening in.)

I don't know how often our guy has his talk button pressed and how far my voice behind him carries, but it's possible that the transcript of this mission would be somewhat irregular:

"Approaching village, over."

"Copy, Liberty. Any update from the target site?"

"You need to put some sunscreen on the back of your neck."

"This is Hammer in the overhead. We have four military-age* males who appear to be orienting themselves to the objective area."

"Copy that, Hammer."

"So do the Taliban use hearing protection?"

"This is Hammer. We've got an exodus of women and children from the village. Two other military-age males messing with something under a tarp."

"Start surging assets."

"Halo, you are approved for rockets and guns, over."

"All these holes in the ground—are they from mortars or, like—"

"Prepare to attack!"

"—gophers?"

"Attack imminent!"

Simulated kinetics ensues. With Mary right behind me, I scramble to stay as close to our guy's back as possible without rear-ending him when he stops to shoot. I try to picture what the group of us must look like, but my brain can't decide between *Zero Dark Thirty* and the Bunny Hop. I imagine officers walking back from lunch, one nudging the other: "What's going on out there?"

---

* In Afghanistan, this means twelve and up, a designation we in the West innocently reserve for toys and board games.

"Audiologists."

The mission ends back by the classroom. We turn in our gear and head inside for a Q&A session with the Special Operations men. They sit in mismatched office chairs in a row at the front of the room. "How many of you," the first question goes, "have hearing loss?" All twelve raise a hand. By one (pre-TCAPS) study, Special Operators, as they are called, had the highest rates of hearing loss in the Army. Both in training and on the job, they spend a greater than average amount of time around explosives and large, noisy artillery. Unless they're snipers. They're either very loud or very quiet, these men.

"I don't understand," says a voice from the back row. "As an audiologist, I never have people come in to my clinic going, 'Oh, my god, I can't hear! I had an incident, and now my hearing is diminished.'"

Chair number 8 explains: "Guys want to go back in and do the job." If a hearing test turns up a loss in excess of a prescribed amount, it can mean being declared unfit for duty or having to secure a waiver to get around it. These are men who, by and large, love what they do. They avoid audiologists for the same reason they avoid doctors.

"I don't want to stop doing what I'm doing," agrees chair 3. "When I take those tests. . . . How can I say this? I want to pass. So I'm like, 'Okay, I *think* I hear a tone.'" Cheater!

Also? This is Special Operations. *Oh, my god, I can't hear!* is not in the script. When things go kinetic, there's a greater than 50 percent chance that a member of the team will be injured or killed. Hearing loss isn't something they spend time worrying about. It's a given. "You expect," adds chair 2, "that you're going to take some kind of degraded hearing on separation." Fallon told us that as an artilleryman, he *wanted* a hearing loss, because everyone in his unit had a hearing loss. "If you didn't have a hearing loss, that meant you hadn't done anything." It might also mean you were born with a robust medialolivocochlear (MO) reflex, which directs the brain to lower the volume on egregiously loud sounds. Nature's TCAPS. Naval Submarine Medical Research Laboratory researcher Lynne Marshall, who is here today, has been working to develop a simple test to identify people with weak MO reflexes so they can be given extra protection.

Chair 6 chimes in: "They're pushing TCAPS for, like, *Hey, protect your ears*. But for us the main function is the comms. The situational awareness." According to a Hearing Center of Excellence fact sheet, 50 to 60 percent of one's situational awareness comes from hearing.

Fallon calls for one last question before we leave for dinner. Again, it comes from the back row. It's almost more of a plea: "Has an audiologist *ever* done anything positive for *any* of you?"

"Yes," volunteers chair 5, a dark-haired, dark-eyed, just generally dark sort who hasn't said much until now. "They fitted me for my hearing aids."

Whomp, wha? Virile, omnipotent Special Ops man wears hearing aids? My reaction is the same mildly stunned one I had upon reading that Angelina Jolie had had her breasts removed. The man went on to question the policy of declaring someone like him unfit for duty. "We let people have devices for corrective vision. Well, I have a device that helps my hearing." What's the difference? It occurs to me that the US Special Operations Command may succeed at something perhaps more challenging than killing Osama bin Laden: erasing the stigma of hearing aids.

I T IS an interesting fact that retired four-star general David Petraeus was shot in the chest on a firing range but not, at the moment, a comforting one. Not that it's Craig Blasingame's job to be comforting.

His job here at the Camp Pendleton firing range, and he's doing it nicely, is to knock out of us any complacence that might be lingering after the run-through of the nearest helicopter medevac points and what to do if searing hot bullet fragments fly down the back of our shirt while we're firing our semiautomatic M16A4 assault rifle. ("Just say, 'Hey, I got some brass.'")

The Special Ops guys will be serving as our shooting tutors. We'll be firing two magazines of ammo each, one with earplugs, one with TCAPS. Ostensibly, this is to demonstrate how hard it is to hear commands while shooting with passive hearing protection in place. It was also, I'm guessing, audiologist bait: *Come shoot M16s with the men of Special Operations!* (Worked on me.)

Craig splits us into two groups, half on the firing line and the rest, including me, a few yards back in the ready box. "Now if this isn't for you," Craig is saying, "if you start to freak out, you can put your weapon down, put your hand up, and say, 'This isn't for me.'" If only war were like that.

To get an earplug far enough in to do its job, the pinna—part of the outer ear—must be pulled out and back, an impossible task while wearing a combat helmet. No one, in the heat of a firefight, is going to pause to take off her helmet, pull back her ear, insert the plug, and repeat the whole process on the other side, and then restrap the helmet. There's time for this on a firing range, and there might have been time on a Civil War battlefield, where soldiers got into formation before the call to charge. Back then, or out here, you knew when the mayhem was about to start, and you had time to prepare, whether that meant affixing bayonets or messing with foamies.

There's no linear battlefield any more. The front line is everywhere. IEDs go off and things go kinetic with no warning. To protect your hearing using earplugs, you'd have to leave them in for entire thirteen-hour patrols where, 95 percent of the time, nothing loud is happening. No one does that. That's why Fallon says, "The military doesn't have a noise problem. It has a quiet problem."

"Group 2," yells Craig. That's me. "Advance to the firing line!"

"Hey, how are you?" says my instructor. "My name's Jack." Jack is unlike the Special Ops guys I have met elsewhere. He's friendly as a Labrador retriever, clean-shaven as a regional sales manager. Perhaps he's carrying out covert ops in San Diego or Scottsdale and, like the bearded Special Operators in al-Qaeda country, needs to blend in with the local male populace. Perhaps he's between missions.

Jack points to my helmet. "Those straps need to go over your ear cuffs. Now that's going to make your helmet tighter, so you probably need to loosen them a little bit." One of the problems with over-the-ear TCAPS is not the equipment per se but the order in which gear is distributed. Helmet fittings used to happen before TCAPS gear was handed out. Guys would try to put their helmet on with the TCAPS headset and now it would be too tight. This seemingly minor planning boner has cost a lot of men a lot of hearing. The one time an IED exploded near Jack, he wasn't wearing his TCAPS. "It was hot, and they were giving me a headache, so I opted not to wear them on that one patrol. And that was the one I got blown up on and had significant hearing loss. Aaron had the same thing."

To my right, an extremely lethal hearing professional has already emptied his first magazine. I'm still battling my helmet straps. "Let me help you," Jack says. I drop my hands to my lap and let him take over. "Oops, I don't want to pull your hair." The gentle sniper.

Jack passes me the M16. "Have you shot a gun like this before?" I shake my very heavy head. He hands me a magazine and shows me where to load it. I've seen this in movies—the quick slap with the heel of the hand.

Hmm.

"Other way. So the bullets are facing forward."

The M16 has a scope with a small red arrow in the center of the sight. You align the arrow with what or (jeez) whom you wish to shoot and squeeze the trigger. Both "squeeze" and "pull" are exaggerations of the motion applied to this trigger. It's a trivial, tiny movement, the twitch of a dreaming child. So quick and so effortless is it that it's hard for me to associate it with any but the most inconsequential of acts. Flipping a page. Typing an M. Scratching an itch. Ending a life wants a little more muscle.

The crack of an M16 is around 160 decibels. Jack estimates he's fired a hundred thousand rounds in his ten-plus years in Special Operations. Weapons and explosions, rather than ongoing "steady-state" noise from vehicle engines and rotors (and MP3 players),* are the biggest contributors to the $1 billion a year the Veterans Administration spends on hearing loss and tinnitus.

Most of those hundred thousand rounds may not even have registered, not because Jack had hearing protection on but because his attention was elsewhere. "When you get in a gun fight and you're up close and personal," he says, "your mind triages what's most important to you." It's a survival mechanism, called auditory exclusion. The possibility that you may lose a little hearing doesn't make the cut.

A sniper also doesn't, I'm guessing, pay much mind to the kind of thing I'm focused on right now: that raising your arms to hold a rifle while lying on your belly causes your ballistic vest to ride up

---

* According to the Department of Defense Hearing Center of Excellence, 12 to 16 percent of American children ages six to nineteen have noise-induced hearing loss. And not from vacuuming and mowing the lawn. Full volume on an MP3 player is 112 decibels, enough to cause hearing loss after one minute. Have you seen Die Antwoord live? (120–130 decibels.) I'm sorry for your loss.

and hit the back of your helmet, tilting it down over your forehead so that it pushes on your eye protection, causing the lenses to knife into your cheeks.

"How do you *do* this job?" The petulant writer. Jack doesn't answer for a moment. He must get this question a fair amount, and most of the people asking are not thinking about the aggravations of incompatible ballistic protection items.

"There's a lot to get used to."

IMAGINE THE Special Operators were paid for their time today, but it's also possible they did it for the steak. The Camp Pendleton catering staff have placed in front of Jack and myself a filet mignon the size of a grenade. Fallon got the fish. He looks like he's about to cry.

"You know what the hardest thing for us is?"* Jack glances around the table. "This right here."

"Yeah." I get it. Strangers with their questions and assumptions. It turns out Jack wasn't referring to any of that. By "us" he didn't mean snipers or Special Operators. He meant the hard of hearing. And "this right here" meant a loud dinner table. Jack says some of his peers cope by asking a lot of questions and pretending to hear the answers. "You see them sitting there nodding, going, 'Uh huh, uh huh.'" Others just withdraw from the interaction.

A version of this withdrawal happens in combat. I tell Jack and Fallon about the work of a team of researchers with Walter Reed's

---

* Apparently nothing. In 2008, a team of psychologists asked nineteen snipers who had served in Afghanistan what they'd found most troubling. Ninety to 95 percent reported having little or no trouble with killing an enemy, handling or uncovering human remains, engaging in hand-to-hand combat, being wounded, having a buddy shot nearby, or "seeing dead Canadians." (It was a Canadian study.)

National Military Audiology and Speech Center. Doug Brungart and Ben Sheffield have been documenting the effects of hearing loss on lethality and survivability. (Because the data-gathering requires Sheffield, with his clipboard, to run around in the midst of the action, military exercises stand in for actual combat.) Members of the 101st Airborne Division agreed to wear special helmets rigged with hearing loss simulators. Among the top-performing teams, even mild hearing loss caused a 50 percent decrease in "kill ratio" (the number of enemies eliminated divided by the number of surviving teammates). Not so much because their difficulty hearing was causing them to shoot or run in the wrong direction, but because they were unsure of what was going on. With their ability to communicate compromised, their actions were more tentative.

Withdrawal carries over to the home front. Brungart told me about a Marine he'd worked with who had lost an arm and a leg and ruptured both eardrums in a blast. "He told me far and away the worst of the injuries was the hearing loss, because he couldn't communicate with his wife and kids." Despite or possibly because of their low profile, the less visible injuries of war can be the hardest kind to have.

4

Below the Belt

The cruelest shot of all

THE AMPUTEES WEAR SHORTS. I see them crossing the Walter Reed lobby, chatting with the security guy, standing in line at this or that café. It's not shorts weather. It's December 4, in Maryland. Christmas music ever in the background—jingle bells, holly jolly, Frank Sinatra agitating for snow. While it is true that a prosthetic leg is immune to the cold, this baring of limbs is about something else, I think. It's an avowal of normalcy, of moving through the world with your hardware on show, no self-consciousness, no big deal. The era of the sad, stiff flesh-tone appendage is over.

Between a man's carbon-fiber, vertical shock-absorbing, micro-processor-controlled prosthetics, it's another matter. You don't hear much about the injuries collectively known as urotrauma, or the techniques used to deal with them. Partly it's the numbers: 300 genito-urological patients for 18,000 limb amputees. It's not that insurgents don't make big enough bombs. It's that bombs that big create corpses, not patients. Advances in combat casualty care, swifter medevacs, and field hospitals closer to the action have meant that more men are surviving who need genital reconstruction. The work remains relatively low-profile, though, because genitals themselves are low-profile.

The clocks on the lobby wall say it's 9:00 a.m. here in Bethesda (and 6:00 a.m. in Los Angeles, and midnight in Guam). I've been passing time in a café before heading up to Urology. A Navy officer practices his Spanish on a woman refilling the condiments caddy. "Thank God it's *viernes*!" A stooped veteran looks at CNN—an Emirates airliner blown sideways during takeoff. "I've done that before," he says to no one specific. Walter Reed is officially categorized as a national military medical center, but it has more the feel of a small indoor town. The larger corridors have been given names: Liberty Lane, Heroes Way, a Main Street with a post office and some fast food outlets. A poster board propped on an easel outside Dunkin' Donuts announces that Colin Powell is doing a book signing at 11:00 a.m.

While General Powell is putting a Sharpie to the pages of *It Worked for Me*, while Guam sleeps, Gavin Kent White will be having his urethra rebuilt. Captain White, a 2011 graduate of West Point, stepped on an IED in Afghanistan. *It Didn't Work as Well for Him*.

THEY ARE buried in twos and threes: one IED to kill the people in the vehicle, the others to kill the people who come to help. White saw the first blast from his lookout in the command and control vehicle on a route clearance mission on a heavily booby-trapped stretch of road in Kandahar Province. He was leading a platoon of combat engineers—specialists in construction and demolition: roads, walls, bunkers, bridges. A Humvee carrying Afghan National Army soldiers, partners of the US and NATO in the conflict, had ignored

White's warning not to drive on ahead. Three were killed, three wounded. The vehicle landed on its side, blocking the road, and it fell to the engineers to move it. White's footstep on a buried pressure plate set off the second explosion—a twenty-pound "victim-operated" IED. I asked him what he remembers.

White lies in a hospital bed, propped against pillows but on top of the bedclothes, on the fourth floor of Walter Reed. The view is impressive, but after four months, you imagine he's fairly well through with it. It began, he says, with intense red-orange in his field of vision and a feeling of lifting into the air. "I sat up, took out my tourniquet, and put it on my right leg, which I saw was missing." The full length of White's other leg remains, but the calf was blown off. He was unaware of this at the time. Because his boot and the front of his pant leg were intact, he assumed the leg was, too.

You sometimes hear that the first words of a man in White's situation go essentially like this: *Is my junk okay?* White's first concern was his soldiers: Was anyone bleeding to death? "I started calling out, 'Who's hit? Who's hit?'" White was their commander, but any soldier's first thoughts, post-explosion, are likely to be of fellow soldiers. Walter Reed surgeon Rob Dean, a colonel who served in Iraq, confirmed this. "The first thing they ask is, 'Where's my buddy? Is he okay?'" Which could, I supposed aloud, be a reference to one's penis. "No," Dean said. "Because the second thing they say is, 'Is my penis there?'"

Despite the assurances of the medic ("Everyone's fine, sir; it's just you"), despite the fact that one leg was maimed and the other was elsewhere, White kept trying to get up to check on his soldiers. Appraise the situation. Be the commander. The medic had to strap him down. For better or worse, this kept him from taking more detailed stock of his injuries. In the immediate aftermath, he had seen that the tip of his penis was "flowered out" but was unsure how deep the damage went. (The verb *to flower* has found an incongruous home in descriptions of IED injuries. In the typical underfoot blast, leg muscle is blown out away from bone, and into that open bloom shoots a dense, fast-moving cloud of bacteria-laden dirt. The blossom then closes over the soil, making the wound hard to clean and prone to stubborn infection.)

White would have thirty-nine minutes to think about it. That's how long the medevac helicopter took to arrive. "At one point I was like, 'If my dick is gone, just leave me here.' I was half-serious. I don't have any kids yet. I didn't want to have to go back without anything to do that with." His men tried to reassure him. "They were like, 'Your dick is fine, sir.'" I'm guessing that that's White and his soldiers right there, in those five words: The formality and respect of "sir" with the easy slang of "your dick."

"I was like, 'Bullshit, I saw it. I just want to know, Is it fixable?'"

It's fixable. Some urethral scarring and tightening has slowed urination and created some erectile torque, but surgery this week should remedy both, as well as some minor cosmetic damage.

Though the pain was heavy enough that White asked a medic for a second dose of fentanyl ("I can't, sir; you'll die"), he has little to say about it. "Honestly, I was more focused on my soldiers." Though they were physically unharmed, a kind of psychic unraveling occurs when a leader falls. White could see how shaken they were, and tried to joke around with them: "Guess my running career is over, heh. Never really was any good at it."

It's hard for me to imagine: worrying about the emotional state of other people when you yourself have just lost part of both legs and possibly some of your genitalia and on top of that your pelvis is broken. White told me his platoon sergeant said to him recently,

"Maybe it happened to you because you're the kind of person who's tough enough to handle it." I think White is plenty tough, but I don't think we're talking about toughness here. This is some kind of blinding selflessness, the sort of instinct that sends parents running into burning buildings. The bonding of combat, the uncalculating instinct of duty to one's charges and fellow fighters, these are things that I, as an outsider, can never really understand.

I emailed White the night after we met. It began as a thank-you, but came around to a sort of grasping fan letter. My world is full of people, and that includes me, who never have to put their lives and bodies on the line for other human beings or for things they believe in. *Hero* has always been a movie word, a swelling orchestral soundtrack word. A Walter Reed hallway word. Now it has something under it.

SURGERY PATIENTS are announced like guests at a ball. An orderly wheels them in and recites from the paperwork: name, age, procedure, body part. To be sure the surgeons are in the right room, with the right patient and the right piece. In White's case, you might otherwise wonder. A nurse is swabbing the surgical site, applying the standard antiseptic man-tan, but she's at his face, not his groin. Major Molly Williams, the almost comically pregnant assisting-surgeon, explains that a replacement stretch of urethra will be built from a strip of tissue harvested from the inside of White's cheek. Mouth tissue makes an excellent urethral stand-in. For one thing, it's hairless. Urine contains minerals that, were there hair growing in your urethra, would build up on the strands. The stony deposits are troublemakers, obstructing flow or breaking free and getting peed out in a blaze of pain.

The surgeon, James Jezior, has been over at the scrub sink going at his nails. He joins us now, hands front, drying. He has blue eyes and fine sandy hair and a mischievous wit. I would use the adjective *boyish*, but on paper he is very much not a boy. He's a chief (of the Walter Reed urology department), a director (of reconstructive urology), and a colonel.

"Also," says Jezior, "the mouth is tolerant of pee." He means that the mouth is built for moisture. It's possible to create a urethra from hairless skin on the underside of the forearm or behind the ear, but the frequent wetting from urine can degrade it. A kind of internal diaper rash may ensue. Inflammation eats away at the tissue, tunneling an alternate path for the waste, called a fistula. Now you are dribbling tinkle from a raw hole in your skin. Just what you need.

White's face has been draped with a blue sterile cloth with a single opening, reminiscent of an Afghan burqa. In this case, the opening is positioned over the mouth, not the eyes, as though the patient belonged to some esoteric spin-off sect. Retractors square White's mouth, pulling it wide to either side, the way kids will do with their fingers to frame a stuck-out tongue. Jezior outlines the graft with a surgical marker and uses an electrocauterizing tool to cut it free. A vaguely familiar aroma, somewhere between brazier and burning hair, hits the air. Jezior is indifferent to it but reveals that the prostate, when cut open, releases a distinctive scent that's kind of nice.

Using long-handled forceps, Jezior passes the dangling tissue to Molly. They look like a couple sharing a Chinese entrée. Molly drapes the graft over one gloved thumb and, with her other hand, snips away bits of fat and tissue to make it thinner. It takes time for new blood vessels to grow in and service a graft. For the first couple of days, the cells of the graft are nourished by a broth of serum. If the graft is too thick, only the cells on the surface will thrive, and those on the interior will die. For this reason, larger skin grafts, like the ones on the back of White's remaining leg, are run through a mesher. The holes of the mesh create more surface area for the business interactions of cellular life: nutrients in, waste out.

If replacing part of the urethra doesn't resolve the problem, another option would be perineal urethrostomy. Here the surgeon would excise the damaged portion and thread the shortened urethra through an opening in the perineum—the no-man's-land between scrotum and rectum. "Then they have to sit to urinate, like ladies do," says Molly.

How big of a deal is that? Jezior makes the point that someone whose reproductive organs have been damaged by an IED has typically also lost one or more limbs. Having to sit down to urinate probably doesn't rank high on the worry list.

Molly tilts her head to face me. "It's huge." Depending, to some extent, on culture. Some years back, she attended a session on perineal urethrostomy at an international urology conference. The Italian surgeons were aghast. "You can't tell an Italian man he's going to have to pee sitting down."

Molly was one of two female urologists at the meeting. She notices the disparity, but it doesn't faze her. On the upside, she never waits for a toilet during session breaks. "I've been the only one in the women's room at some of these urology conventions."

"Same here," deadpans Jezior.

The piece of cheek is ready to begin its new career. A nurse pulls a sterile drape from White's hips and begins rubbing his skin with the antiseptic wand. Such is the vigor of the youthful male that even under general anesthesia, even when it's a ChloraPrep sponge bestowing the caress, the penis responds. It is a less robust response than normal, perhaps, because Jezior has prescribed something to temporarily blunt erections. Surgical incisions are sewn up while the organ is flaccid; erections stretch the incision. They *hurt*. However, erections bring more blood into the penis, which speeds healing, and they also help prevent scarring. The latter is important because scarring—especially in erectile tissue—can make erections crooked and uncomfortable. For this reason, sexual activity is sometimes encouraged postoperatively as a kind of physical therapy for the penis. Walter Reed nurse manager Christine DesLauriers, whom we'll shortly meet, convinced the intensive care unit staff to establish a daily "intimate hour," during which no medical staff would visit the patient's room, just spouses and partners.

Jezior opens the organ to access the urethra. As he works, he rests the heel of one hand on White's scrotum, using it like a tiny beanbag chair. Molly's style is more formal; she holds her instruments like a knife and fork, wrists raised. The rectangular graft is stitched in place but left flat. Urine is temporarily diverted through an opening made in the skin below the graft. In a follow-up operation, once a new blood supply grows in and it's clear the graft has taken, Jezior will go back in and hook up the waterworks. He'll roll the graft into a tube and connect it to the original urethra, and that, one hopes, will be that.

When it's over, Jezior snaps off his gloves and walks directly to a phone on a desk in the corner of the operating room and punches an extension. White's mother is waiting in his hospital room. "He's awake, and everything went well."

FOR THE third time today, I've lost Dr. Jezior. I'll bend down to slip on some surgical shoe covers or step away to use a drinking fountain, and when I turn back he's gone: pulled away by a nurse, an administrator, a patient's wife. He never says no, although he has every reason to. Chronically over-busy, he moves through the halls at a slight forward cant, as if arriving a second sooner might give him a jump on the enduring backlog of things that need doing. The stack of reading material in his office bathroom, all of it urological, threatens to collapse the sink.

Like a lost child in a mall, I know to stay put and eventually he'll come for me. I browse some information on "Boxes and Storage," one of the many themed bulletin boards that line the corridors of Walter Reed. "Mature Indian wheat moth larvae pupating in corrugated cardboard," says a photo caption. It's the most unsettling image I've seen all day, but not for long. Jezior and I are headed to his office so he can show me photographs of some of his patients in Iraq. Not to unsettle me, but to give me a broader sense of what bullets and bombs, and then surgeons, can do.

Jezior narrates with simple anatomical vocabulary, but I can't always parse what I'm seeing in a way that matches the words. I can't even see *person* in some of these images. I see *butcher shop*. Bandages protect the psyche, too; some of these soldiers never saw what I'm seeing. Jezior had a patient who didn't see the injuries to his penis for three weeks. He clicks ahead to a slide from this man's arrival at the hospital, a close-up of the weapon-target interaction, as they say in ballistics circles. How do you prepare a patient like this for the unveiling? "We used to try to sound optimistic," Jezior says. "But when this guy finally saw it, he was like, 'Oh, my God.' It was another devastation, a second loss." Now they're blunter. "I'll say, 'It's a severe injury. You'll have to see it.'" If there's going to be a surprise, let it be a positive one.

What can be done for these men? A lot. The art of phalloplasty—crafting a working penis from other parts of a patient's body—has come a long way (thanks in no small part to the transgender community).

To build a penis, Jezior begins with an arm. A rectangular flap of skin on the underside of the forearm is planed into two thinner layers. The inner one is rolled to form a urethra; the outer becomes the shaft. This tube within a tube is left in place, nourished by the arm's blood supply. When what remains of the original organ heals, the new model is detached from the arm and reattached farther south.

Erectile tissue is the challenge. While spongiform erectile tissue exists in other parts of the male anatomy—along the urethra and in the sinus cavity (congestion being an erection of the nasal turbinates)—there isn't much of it, and no one has tried to transplant it. And while there are eye banks and sperm banks and brain banks, no one is banking noses. So in place of the corpora cavernosa—the two parallel cylinders of erectile tissue—surgeons install a pair of inflatable silicone implants. (To get erect, the patient—or his friend—squeezes a little silicone bulb implanted in the scrotum that pumps saline from a receptacle in the bladder.) Hook up the tubes and let the nerves regrow, and in time orgasm and ejaculation are back on track.

Jezior continues with his slides. "This is a brigade commander. A sniper shot him across the top of the groin. Took out the middle part of his penis." Losing the whole penis—and surviving the blast—is rare. Among Grade 3 and higher (the worst) cases of Dismounted Complex Blast Injury, 20 percent suffer damage to the penis, but only 4 percent lose everything.

You have to wonder: Was the sniper off his game, or was the shot intentional? Are there some who aim for the crotch? Jezior thinks that there are. He's heard stories from World War II. Dale C. Smith, a professor of military medicine and history at the nearby Uniformed Services University of the Health Sciences (USUHS), has also heard those stories, but knows of no evidence to back them up. Smith points out that the secondary goal of a sniper is to sow

fear. In that sense, the crotch is an effective shot. However, Smith said in an email, it is also a risky shot, in that a sniper is looking for a "high percentage return" on the tactical effort and risk of getting into position. The pelvis is not considered a "kill shot."

Another gunshot case follows, this one through the scrotum and rectum. "This is half his anus here. Here's his scrotum up here. This is the insides of the testes. " The horrid Cubism of modern warfare. The reconstruction in this case was done by Rob Dean, Walter Reed's director of andrology. The andrologist's beat is reproduction, not excretion: testes and scrotums, hormones and fertility. Dean is joining Jezior and me in a few minutes for lunch, in a sandwich place downstairs. The two served four months together in Iraq.

Jezior closes the photo file and leads me out through the urology waiting area, toward the stairs. "Patient Jackson?" calls a receptionist. As though "patient" were the man's rank. I guess in a sense it is. He may be a major or a colonel and the man across from him may be a private, but here everyone's a patient. In a culture defined by rank and hierarchy, Walter Reed can seem—to an outsider, anyway—endearingly egalitarian.

Dean is already in the line to order sandwiches. He, too, is extremely busy, which, in the grand and ghastly scheme of war, is a good thing. It means more men are surviving bigger explosions. If funding and research lag behind, it's partly because of the general cultural discomfort that surrounds all things sexual—including the poor organs themselves. On a much simpler level, Jezior says, it's a case of out of sight, out of mind. "When some celebrity comes to Walter Reed and visits you in your room . . ."

Dean jumps in. They finish each other's thoughts like an old married couple. ". . . Right, the President doesn't pull down the sheet and go . . ."

". . . 'That's terrible, look at that. His penis is gone. Let's get some money flowing for that.'"

Walter Reed Medical Center pays for phalloplasty, although there was initially some resistance. (The implants alone cost about $10,000.) Erections were thought of as "icing on the cake," Dean says. "They'd say, 'Oh, people don't really need that.' I'm like, 'Well, the guy with the amputated legs doesn't *need* prostheses. Put him in a wheelchair!' And they'd go, 'Oh, no! It's important that they walk!' I'd say, 'Okay, well, most people think it's important to have sex.' Can I get a Caprese sandwich and a Coke Zero?"

Dean has expressive hands and eyes and prominent arching eyebrows, and when he talks and laughs, the whole lot of them join the fun. In this business, humor and candor are a therapy on their own. Dean has been known to put a ruler to a discouraged patient's penis and hoot, "You've got six inches! How much more do you need?"

Don't be fooled by the jolly tone. Dean is a bulldog for his patients. He was a force behind the push to get the VA to cover in vitro fertilization for soldiers whose injuries left them sterile. He gives talks to USUHS students about sexual health issues among injured service members and answers questions at veterans support groups. He helped colleague Christine DesLauriers found the Walter Reed Sexual Health and Intimacy Workgroup: a dozen-plus local medical providers and social workers who gather periodically to plot strategy and share resources. For instance: *Sex and Intimacy for Wounded Veterans*, a book by DC-area occupational therapists Kathryn Ellis and Caitlin Dennison. These two do not flinch. Here are sexual positioning tips for triple amputees. Ways to modify a vibrator for a patient who's lost both arms below the elbow. I second the sentiments of the title page endorsement (if not the precise

phrasing): "We should put a copy of this manual in the hands of every patient, spouse, and medical provider . . ."

Especially the medical providers. "It's amazing," says DesLauriers, "how many of them are frightened to bring it up." She told me about a Marine she'd worked with who said to her, "Christine, I've had thirty-six surgeries on my penis, I've had my shaft completely reconstructed, and not one damn person told me how I'm going to go home and use the thing on my wife."

Few talk to the wives, either. "It's depressing watching some of them interact," says Jezior. "In your mind you're going, 'She's going to leave him.'" When I asked DesLauriers what the divorce rate is, she said, "Divorce rate? How about suicide rate. And what a shame to lose them after they've made it back. We keep them alive, but we don't teach them how to live." Walter Reed has no full-time sex educators or sex therapists on its payroll. The Internal Medicine Clinic offers appointments in "sexual health and intimacy," but only one nurse is set up to handle them.

"It's not," Jezior says when the topic comes up, "as well situated as we'd like it to be . . ."

Dean cuts through it. "There's nothing. There's a vacuum."

DesLauriers' workgroup has spent seven years meeting with military boards, trying to get Defense Department funding for an on-staff sex therapist at Walter Reed. She gets lots of support, almost entirely verbal. The problem isn't just budget cuts. "The problem is getting the US government to embrace sex." She told me about a meeting several years ago with an admiral who headed up Walter Reed. "He said, 'I don't understand what we are teaching someone who doesn't have a penis. What exactly are you going to help that person with?'"

There are so many things DesLauriers could have said to the admiral. She could have said, "Strap-ons, sir? Thigh riders?" She could have quoted from Ellis and Dennison's book. "'Incorporation of a residual limb in creative ways, such as stimulating a female partner's clitoris,' sir?" "'Exploration of other areas that could provide more pleasure (e.g., nipples, neck, ears, prostate, rectum),' sir?" She went with something more basic: "I said, 'Sir, if I can be very candid with you. Does he have a tongue, and can he be taught?'"

"The other thing to keep in mind," Jezior says, "is that in the early stages after a major injury, there's a lot going on that makes sexual intimacy not necessarily the priority . . ."

Dean, nodding: "Like, *Can I brush my own teeth now?*"

"And they're heavily medicated to get them through this period." Narcotics, nerve stabilizers, antidepressants. "So if they're not getting a good erection, you say, 'Let's get you through this, get you off the pain meds, and then see how you're doing.'"

Or, if you're Christine DesLauriers, you say, "Can you handle a bit of pain? Cut back on the meds for four hours, have sex, go back on the meds." Catheter in the way? Fold it back and put on a condom. "Absolutely you can have sex with an indwelling catheter!"

Aside from Christine DesLauriers, are there other promising developments? What's on the urotrauma horizon? What about penis transplants? I'm only half-serious, but Jezior starts talking about experimental work going on at Johns Hopkins.

"Wait, they're going to transplant a penis?" Some extraneous decibels on that. A couple look up from their paninis.

Jezior says, "Yeah"—the kind of *yeah* you give someone who's asked if you want your receipt, or fries with that, like it's nothing. He adds that one of the patients in the photographs we were looking at is a candidate. Though it won't happen for at least six months. "They're doing some cadaver work right now."

"Really."

5

It Could Get Weird

A salute to genital transplants

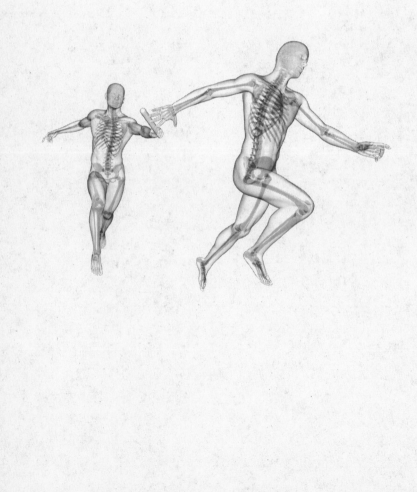

THE ELDERLY DEAD—THE MEN, anyway—always seem to need a shave. Maybe it's because their demise so often unfolds over a span of days. While dying leaves plenty of unscheduled time one could use for shaving, for trimming one's toenails or arranging one's hair, there is little energy for sprucing up and really no call. The two dead men lying on gurneys in the cadaver lab of the Maryland State Anatomy Board this morning share the look—stubble and bed hair—but aside from that, they appear nothing alike. One is fleshy and barrel-chested. His legs are splayed at the hip with knees bent, one higher than the other. The carefree legs of a man dancing a jig. The other cadaver is rigid and lean. His legs lie pressed together like chopsticks. You could almost slide him under a teller window. One body has a tattoo, the other has none.

One is circumcised, and one is not. Given that the surgery being worked out this morning is a penis transplant—a lead-up to the first such operation in the United States—this is the difference that stands out. Though of course it doesn't matter. The recipient will never wake to see his new endowment. Thus the cadavers weren't chosen for any particular genital attribute. "They are whoever happened to be on hand," says Rick Redett, the surgeon heading up the session, "and male."

Redett and the plastic and reconstructive surgeons assisting him—Damon Cooney and Sami Tuffaha—are from down the road, at Johns Hopkins University. The Hopkins School of Medicine, with funding from the Defense Department, has been the setting for a lot of innovation in the field of transplantation over the past decade. The members of the surgical team that performed the first double-hand and the first above-elbow transplant in the United States are there now. Hopkins transplanters helped refine a technique called marrow infusion, which greatly reduces the likelihood that a patient's body will reject its new parts. This is especially helpful with transplants of composite tissue. A face or hand—unlike a liver or kidney—is a variety pack of skin, muscle, mucous membrane. If you're talking about a penis, add erectile tissue to the list. The body may accept one or two kinds of tissue and reject another. Skin is especially problematic because it's a protective barrier; immunologically, it's on high alert. To fool the body's sentries, patients receive an infusion of the donor's bone marrow—marrow being a generator of immune cells. The donor's marrow doesn't replace the patient's own, but it reprograms the immune agenda to an extent. The body may glower suspiciously at its new parts but stops short of wholesale eviction. A lower risk of rejection means fewer immune-suppressant drugs are needed, and at lower doses. That, in turn, means fewer side effects and healthier patients.

New techniques like marrow infusion have tipped the ethical balance for transplants that are non-life-saving. The benefits of a face or hand—and maybe a penis—transplant have begun to outweigh the drawbacks. (Legs are a less appealing type of transplant, partly because the nerves have so far to regrow. For now, prosthetics present a better option.)

Redett heads the Johns Hopkins transplant team's reconstructive and plastic surgery arm, and, like me writing this sentence, will stick a body part most anywhere. Earlier he described separating a set of conjoined twins. The sentence ran like this: ". . . so we transplanted the dying sister's leg and buttocks and a little bit of her pelvis and then we took her aorta and plugged it into . . ." Redett's own features are solidly After-photo: the face well balanced, the nose small to average-sized, the eyes pleasingly spaced. His voice is the stand-out element. He sounds just like the actor James Spader.

Redett pulls on a surgical cap cut like a knight's chain mail: all the way down over the ears and low across the forehead—the better to ward off cadaver lab smell. (He has a lunch meeting.) Cooney's cap is a bright green luck-of-the-Irish clover-print number that belonged to his dad. Flashes of gray hair can be seen below it, at his temples, though you would not use the word *distinguished* to describe him. *Adorable* you might use. He is forty but looks thirty. He also, in tribute, wears the old man's magnifying loupes, which are too big for his face and keep sliding down his nose. Today he has a cold, well timed given the odors of the morning.

Veterans from Walter Reed often come to Johns Hopkins for phalloplasty—a penis reconstruction made from a cannoli roll of their own forearm skin implanted with saline-inflatable rods. The resulting "neopenis" is impressively natural looking. It is a testament to Redett's skill that some of the pictures on his phone could be mistaken for Anthony Wiener–style selfies.

"This is a soldier who was hit with an RPG in Afghanistan. Lost his testes and scrotum and penis. There's the flap being raised on his arm." Redett swipes through photos like a proud parent. "We made a scrotum using a tissue expander in his perineum. Here it is with the artificial testes. He has total sensation now." After nine months to a year, a patient's penile nerves regrow in the tissue formerly known as arm, supplying normal penile sensations and triggering orgasm very much as they used to.

So why would a man opt for a transplant? Especially since transplants still—even with the marrow infusion—require some degree of immunosuppression. And not only does immunosuppression diminish the body's defenses, opening the door to infections and cancers, but the drugs it requires have hefty side effects. Why not stick with phalloplasty?

"Here's the problem." Redett steps over to a whiteboard on the wall and draws a penis. For a moment, it looks like fifth graders had the run of the place. The problem is extrusion: implants poking through the tip of the penis, typically during intercourse. Penile implants were designed for men with erectile dysfunction (severe cases that Cialis won't help). In these men, the inflatable rods are inserted into tough fibrous sheaths that line the erectile chambers (two of which run the length of the shaft like the barrels of a gun). Phalloplasty patients have no sheaths, just skin—which is easier to poke through. Think of holding a restaurant drinking straw in your fist and pulling down the wrapper until the straw pushes out the top. It's that kind of situation. The extrusion rate has been reported to be as high as 40 percent (though sheathing the implants with Dacron or cadaveric tissue sleeves has helped somewhat). Also, as mentioned in the previous chapter, urethras made from forearm skin sometimes prune up and deteriorate in a moist environment.

Besides, a man might like to have a natural, no-pumping-needed erection. (To get hard, a man with implants has to squeeze a bulb inside the scrotum that pumps saline.) A man might also, when he's finished with that erection, wish to have a less bulky, more retractable organ. Uninflated penile implants are less rigid but no shorter. "Right?"

Cooney glances over his loupes. "In general, Mary? Men don't complain about it being too big."

A S YOU read this, Redett's team may have undertaken their first transplant. When I last checked in, in February 2016, a wounded veteran had been selected and was awaiting a suitable donor. In addition to the matching criteria used with internal organs, a penis must also, Redett said in an email, be a good match visually: "Skin color and . . . age." And size, I wrote back? This he shrewdly ignored.

Their first won't be the world's first. That took place in China in 2006, at the hospital of the Guangzhou Military Command. In the case study, the surgeons describe the recipient not as a soldier but as the victim of an unspecified "unfortunate traumatic accident." Additional trauma ensued: The new penis "regretfully had to be cut off" after two weeks. The man's body didn't reject it, but his wife did. No details were supplied other than to say that there was a "severe psychological problem . . . beyond our and the patient's imagination." Swelling was mentioned, and some necrotic tissue.

Necrosis happens when tissue is deprived of oxygen—in this case, because someone's transplant surgeon didn't hook up the necessary arteries. The skin turns black and leathery and eventually falls off.

"*Necrotic* means dead," explains Cooney. "Surgeons don't like to say *dead*."

Even without necrosis, a transplanted appendage has a taint of death. It's not dead, but it is a bit *resurrected*. You can imagine how a patient might be uncomfortable with it. With internal organs like kidneys or lungs, the psychological consequences are generally mild: out of sight, out of mind. "But it is not so easy to use and see . . . a dead person's hands, nor is it easy to look in a mirror and see a dead person's face," wrote Jean-Michel Dubernard, the surgeon who successfully transplanted the first hand—which was later removed, the patient believing it to be evil. (The hand was swollen and inflamed, though not from evil. The recipient had stopped taking his immunosuppressants.)

Cooney's experience has been otherwise. "People really thought that with the hand and face transplants, conversion"—the psychological assimilation of another person's body part—"was going to be an issue." It has not been. "I realized that that is the whole person's hubris: You and I have two hands, so having another hand would feel unnatural. But having a missing hand is *more* unnatural." Cooney's experience with all six of the hand transplant patients his team has worked on is that the instant they wake up, even though they can't yet feel or even see their new hand, it feels like their own. This has been true even in cases where the hand was from a person of a different gender or with a slightly different skin color.

Receiving a stranger's face has also proved less disturbing than people had imagined, because the alternative is no face at all. "Patients say, 'I don't care whose face I get,'" says Cooney. "Having a face is being human. Not having a face is being some movie monster."

And penises? "I've been trying to think," says Cooney, straightening a row of surgical instruments laid out on the big guy's belly. "What's different about the penis? It's not part of one's identity in the way a face or even hands are. But there's something about it. It's *more* personal, in a way, because no one sees it."

And in this case, everyone will want to. The media spotlight will be intense and especially uncomfortable. "When you've got somebody sitting there in a wheelchair with bilateral arm transplants, it's easy to look at him and say, 'Wow, that is really something,'" says Redett, from his work station at the other gurney. "But when you've got a guy sitting there in a hospital gown, saying, 'Yup, everything went well . . . ,' you know what everyone's thinking: *Does it work? Can we see it?*"

Cooney makes a deep cut, the big man's penis springing open, kielbasa-like, under the blade. When pressed, he will allow that this is, as a male, an uncomfortable act. And then change the topic.

"So this is the spongy tissue of the corpus cavernosum." He indicates one of the twin erectile chambers. He squeezes the stump, and blood appears like water from a sponge.

Because blood is the substance of erection, hooking up the right arteries is doubly important: not only to avoid necrosis, but to facilitate sexual function. The Chinese surgeons didn't reattach the cavernosal arteries, which run down the center of each erectile chamber and supply much of the blood for erections. One reason, perhaps, for the wifely discontent.

Meanwhile, one gurney over, an artery in the skin on the lean cadaver's abdomen is being hooked up to a tube running down from an IV bag. The fluid in the bag is dyed indigo, and when it begins to flow, a patch of skin will blush blue, revealing the precise territory fed by the artery. In this way, Redett and his colleagues are able to pinpoint which vessels are critical for the transplant. There will be no necrosis when the Americans move their first penis.

The IV isn't a drip but a rapid infusion, a setup used in emergency rooms to replenish blood volume quickly. "The first time we tried this, it was a disaster," says Sami Tuffaha, who has been researching penile vasculature as part of his residency. "Dye all over the place." Irritated janitor. Ruined loafers. He sticks out a foot. "They're my cadaver shoes now."

From off behind us comes the voice of James Spader. "If you don't have a pair of cadaver shoes, you're not doing enough research."

In a previous session in the same lab, Tuffaha located a vessel coming off the femoral artery that perfuses the skin of the lower abdomen just above the penis. They're rechecking this, to be sure it wasn't an anomaly. Tuffaha reaches up to open the valve on the IV. Within seconds, a time-lapse bruise unfolds. The area expands and darkens, its boundaries made clear. "This is great," says Redett. "We can take this whole area as part of the transplant." Transplanting a penis is like transplanting a tree. You don't just lop it off at the trunk. You take the ground around it and the roots that nourish it. In all, three to four veins, a like number of arteries, and two nerves will need to be connected.

The donor cadaver, the lean one, lies on his back, one forearm draped across his waist. It's a relaxed pose, a movie pose—postcoital, maybe, or poolside chaise longue. It's an odd visual, given the proceedings. Tuffaha and Redett have by now disconnected the whole package: penis, scrotum, and a peninsula of flesh above and to the side, which contains that critical artery Tuffaha found.

Redett needs photographs for an upcoming conference presentation. Tuffaha obliges by holding the unit in front of the camera. With thumbs and forefingers he dangles it by the two top corners of skin, then reverses it, so Redett can document the back side. Imagine a mother-to-be at a shower, holding up a baby sweater for guests to admire. It's of similar size and floppiness, is what I mean. Possibly there was a better comparison to be made, but let's move on.

I later asked Ronn Wade, who runs Maryland's body donor program out of an office down the hall from the lab, what he would say in the event he was contacted by a family member wanting to know how this cadaver was used. He answered that he would tell them it was a "multi-use clinical/surgical specimen." Having seen what I've seen, I understand the need for vagueness. Before you could expect a body donor's family to accept the specifics of the research under way today, they would need to understand the specifics of its promise. They'd need to have a sense of what it's like to be a soldier or Marine who wakes up from surgery after an IED blasts a hole in his life. They'd need to appreciate that the procedure being developed in this windowless horror movie of a room has the potential to restore the wholeness of a young man: his future, his relationships, his well-being. More graceful, I think, to leave the particulars of the gift unspoken.

THE DONOR'S work is done. Where his penis was,* there's a crimson rectangle, a loincloth of his own tidy gore. The testes, skinned, have been pulled off to the side of the hips. "You're not taking these?" I ask Redett. As though he were packing for a trip. I'm thinking now of combatants whose injuries leave them unable to generate sperm. It might be nice to give them, along with a functioning penis, a reproductive future. What's a few more ducts and tubes to hook up?

It's trouble. That's what it is. Hook up the testes, and now the penis donor is also a sperm donor. If the transplant recipient

---

* In the same way amputees feel phantom pain in the space where the arm or leg once resided, penile amputees sometimes feel phantom pleasure. This, and phantom erections, were first described by the coiner of the phrase "phantom limb," Silas Weir Mitchell. What gave Mitchell his particular expertise? He worked with Civil War amputees at the "Stump Hospital" in downtown Philadelphia.

Oh, for the titular economy of yesteryear. The Stump Hospital is gone and in its place we have the likes of the Veterans Affairs Center of Excellence for Limb Loss Prevention and Prosthetic Engineering. Though all is not lost. We still have a Foot & Ankle Center in London, a Breast Clinic in New Delhi, a Kidney Hospital in Tehran, the Face & Mouth Hospital in Calcutta, New York's Eye and Ear Infirmary, and the Clínica de Vulva in Mexico. The poor penis has no hospital to call its own.

impregnates someone using the dead donor's testes—and, more to the point, his genes—whose offspring will that child be? What if the donor's widow tries to lay claim to her dead husband's sperm, now being generated inside a different man? What if the dead man's parents want a relationship with their biological grandchild? Cooney looks up from the stump: "It could get weird."

I asked Ray Madoff about this. Madoff is a professor at Boston College Law School and the author of *Immortality and the Law*, the go-to book on the legal rights of the dead. "It's no weirder a problem than we already have," she said, meaning that the United States years ago entered the uncharted waters of donor sperm and donor dads. "Some countries, sensible countries, have statutes and regulations about what happens to the sperm of dead men." The United States isn't there yet. It's a place where judges have ordered sperm donors to pay child support, and rapists have been granted visitation rights to a victim's child.

For now, more practical matters stand in the way. It's enough of a challenge to find people who'd be willing to let Rick Redett take the penis from their brain-dead, respirator-oxygenated loved one and stitch it onto another man. Taking the cellular lineage, too, would, as Cooney says, "be beyond the normal donation that most people would consider." In the meantime, simpler options exist. The military could, as a matter of course, bank sperm from each male soldier prior to deployment.

Rob Dean, the Walter Reed andrologist from chapter 4, counters that even that isn't simple. "It's an elective procedure," he said when I visited. "The military can't say, 'Line up, we're going to make you donate sperm.'" There's also a cost-benefit issue. Maybe three hundred veterans from Operation Enduring Freedom suffered injuries that left them infertile. "So for those three hundred you're going to bank sperm for a hundred fifty thousand men?" In the current climate of Defense Department budget cuts, it's a tough sell. Madoff surmised that military budgeteers might have an additional concern. A widow who uses a dead veteran's banked sperm may be creating not just a baby but a government beneficiary.

A third option exists. Sperm typically live about forty-eight hours, so it's possible—if things look testicularly dire—to extract the last batch, the soldier's last shot at biological fatherhood, in the operating room. "But again," said Dean. "If they haven't consented, I can't do it. I don't know if this guy wanted to be a father, now or ever. I need to know that, or have a [prior] directive from a legal guardian or next of kin. The wives and girlfriends get upset, but it's not their body."

So education is what's being done. Information about sperm banks is sent to service members before they deploy, so that at the very least they're aware it's an option.

Not good enough, says Stacy Fidler, a veterans' reproductive rights advocate I spoke to at Walter Reed. With support from a national infertility nonprofit called Resolve, Fidler is pushing for on-base sperm banks. She lives with her son Mark, a Marine who has been recuperating in an apartment at Walter Reed National Military Medical Center since the propulsion charges on three grenades on his belt were set off by a nearby IED. Mark lost all of both legs and both buttocks. Although, quoting Stacy, "the big boy's fine," there was some damage to the testes, and the family doesn't know whether Mark will be fertile once he heals.

MARK WAS on his bed when I arrived. It was midafternoon, and the curtains were closed. *The Big Bang Theory* was playing through a projector set up on his bedside table. I sat down in the one chair available, in the path of the projector's beam. The actors sniped at each other on the side of my head until Mark reached for a remote and shut them down. Pressure sores made it too painful for him to sit upright. Without the cushioning of buttock muscle, the bony points of the pelvis can wear through the skin. Mark's bed had become his couch, his office, and his dining table. Within arm's reach were three remotes, an iPad, a plate of donuts, and that simplest of prostheses, the rattan back scratcher.

"Listen," said Mark. "I know how a grunt's mind works. They're not thinking about having kids. They don't have wives, most of them." He was shirtless under a gray fleece throw, his body a round form that stopped too soon. He pointed out that the sperm bank nearest to the Marine Corps training base at Twentynine Palms was probably in Los Angeles, three hours away. "You can give them all the information you want; they're not going to do it."

His mom joined the conversation. Stacy Fidler wore jeans and a red shirt with a Marine Corps insignia and was perched on the edge of Mark's prone cart, a joystick-operated, wheeled table that he's been using to get around. "It should be available right there on base," she said. "And if you don't want to, you don't have to."

"No," countered Mark. "You have to make them do it. Honestly, in Afghanistan we talked nearly every day about getting blown up. But the most we ever talked about, injury-wise, was losing maybe above the knees, both legs. You never think about the genitals. Don't give them a single chance to go, 'Aaaaa, forget it.'"

If the military were to pay for predeployment sperm-banking for every male recruit, wouldn't they also need to pay for extracting and freezing eggs—a costlier and more involved undertaking? Stacy shakes her head no. "If a girl gets her ovaries blown up, she's not going to be here." Meaning that an explosion that blows up a woman's ovaries is likely to be lethal. "That's a whole different ball game," she said, intending no word play.

Mark has radar for whatever frame of mind a person has brought along into his room: unease, medical detachment, in my case curiosity. With little warning, he rolled onto his belly, pulled the blanket off and slid down the back of his Jockeys. Pointing to where his buttocks used to be, he said, "This right here is my lap." *Was*, he means. His surgeons took skin from the front of his thighs, thighs they were removing anyway, and covered the crater made by the grenades. A dressing as big as a gas cap covered a pressure sore.

Once the sores heal, he said, he wants to try skydiving, horseback riding, calf roping. He wants to act in zombie movies and wrestle alligators. For some reason, it was the next one that made me go gooey: "I want to see Paris." To this day, when I think of Mark, I picture him, cigarette behind one ear, rolling way too fast down Boulevard Saint-Germain.

As I write this, there's been chatter in the media about the ultimate composite-tissue transplant: a whole body. If it were possible to regrow spinal nerves, you could, in theory, sever a soldier's head from his severely mangled body and surgically transplant it—hooking up the arteries, veins, and nerves—onto a freshly decapitated beating-heart cadaver whose tissues have been kept oxygenated via a respirator. A rough version of the procedure was performed by Cleveland surgeon Robert White in the 1960s, using pairs of rhesus monkeys. The heads with their new bodies survived for a few days, paralyzed and unable to breathe on their own. Then rejection issues set in. Vastly better immunosuppressant protocols have brought the whole Frankenstein tale closer to reality, though it still resides in the realm of speculation. Spinal nerves are far more complicated than peripheral nerves. Peripheral nerves, which serve the extremities, are like telephone wires in a sheath. When the wire is cut, the signal stops at that point. But if you reattach the axon it will regrow along the pathway of its sheath. With spinal nerves, the analogy is no longer telephone wires; now you're cutting the wires in a sophisticated computer network. The nerves don't know what they're supposed to reconnect to, which way to regrow, what paths to follow to restore function. The optic nerve is similarly complex. That is one reason no one, not even Rick Redett, has successfully transplanted an eye.[*]

THE THIN cadaver's penis lies on the big cadaver's belly while Cooney finishes isolating nerves and vessels on the stump. They're not going to hook them up this time, because that takes six to ten hours, four to six surgeons, and a microscope on wheels. And was not the point of today's endeavor.

When Cooney is done, Redett picks up the organ and drapes it in place over the larger cadaver's stump. In the way a shopper previews the fit of a shirt by holding it to his shoulders, we have a sense of what this body would look like with the other's penis. Redett steps away to get his camera. I am not preparing a presentation, but I, too, take some photographs. As though I could ever forget the sight.

Redett finishes and sets down his camera. He zips the big man's

---

[*] An exception is made for Dr. H. W. Bradford, who, for cosmetic purposes, transplanted a rabbit eye into the socket of a sailor who'd suffered a childhood eye injury. "The nature of the man's calling," wrote Bradford in the 1885 case study, "made it undesirable to use a glass eye." I don't know the precise occupational risks of the seafaring eyeball, but the prevalence of eye patches among pirates suggests they do exist. Despite some clouding, the operation was deemed a moderate success. Though rabbits have larger pupils, their eyes are otherwise unnervingly similar to our own, as a Google Image search will quickly establish. I can't recommend this activity, however, as the search results will include a photograph of a plastic-lined box captioned "Rabbit heads: no neck, no skin, with eyes. 100 grams each. Please contact me for price quotation."

body bag. It resembles a tuxedo bag and has a space for the cadaver's name, which has been filled in with black marker. When I get back to my hotel, I find an online obituary for him. There are a number of interactive options. One of them causes me to make a strangled barking sound. "Please add a photo and share in the life story of _____." Another option invites me to add a memory to the online guest book. "If you need help finding the right words, view our suggested entries." Nothing seems to fit.

6 Carnage Under Fire

How do combat medics cope?

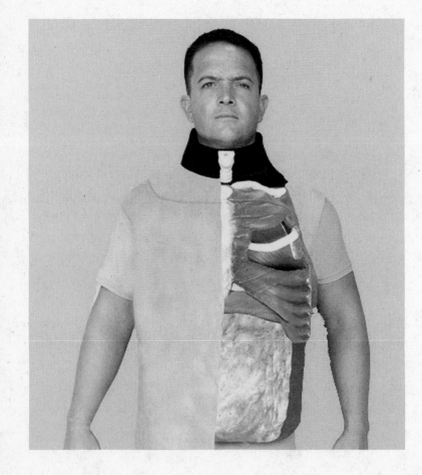

THE CALL TO PRAYER can be heard from the Carl's Jr. parking lot. You can hear it at the Wells Fargo drive-through and outside the offices of the San Diego County Water Authority. The attentive listener will notice that something is off. Rather than five times over the course of a day, you may hear it six or seven times in a morning. Other days it is absent. If, perplexed, you were to follow the sound, you would find yourself not at a mosque but at a spread of movie studios and sets known as Stu Segall Productions. By all means, knock on the door and have a look around.

Segall was born a Stuart, but on his movie credits and in my mind he is always and very much a Stu.* Chest hair can be seen, and some necklace in there. There are whiskers, sparse and longish, somewhere between beard and I-don't-feel-like-shaving. He has a

---

* Except when he's a Godfrey, as he is in many of his 1970s movie credits. Godfrey Daniels produced ten titles in the long-forgotten genre "soft core," paying loving if needless attention to his plots, one of which could be a chapter in a Mary Roach book: "A research facility uses state-of-the-art equipment to test sex dolls."

wife but spends more time in the company of Bob, an agreeable Rottweiler who naps on the black leather couch in his office. Segall dives in and out of careers with glee. Writing, directing, producing (most recognizably, the TV crime drama *Hunter*). He owns a diner next to the studio. He doesn't cook, but occasionally he names menu items, and you can pick them out without too much trouble—for example, the Boob (chicken breast) Sandwich.

Early in 2002, with Hollywood's appetite for action dramas dampened by the events of 9/11, Segall began repurposing his talent for gore and violence. He founded a company, Strategic Operations, to produce loud, stressful, hyper-realistic (the coinage has been trademarked) combat simulations for training military personnel: the fog of war, in a box. Many of the trainees are corpsmen (Navy medics who deploy with Marines and SEALs)—men and women whose job may require them to perform emergency procedures while guns are going off around them and people are screaming and dying and bleeding like garden hoses. The underlying concept is "stress inoculation." If you're thrown into a staged ambush in Stu Segall's Afghan village mock-up, the thinking goes, you'll be calmer and better prepared when the real shit hits overseas. For medics, being calmer matters a lot. The fight-or-flight response is helpful if you're fighting or taking flight but, as we'll see, fairly catastrophic if you're trying to stanch the flow of blood from an artery or cut an emergency airway or just generally think fast and clearly.

Forty future corpsmen for the 1st Marine Division, headquartered in nearby Camp Pendleton, are here today as part of a combat trauma management course. Over the course of two and a half days, the trainees will administer pretend emergency care to role-players, most of them Marines, in six varieties of military pandemonium, beginning with an 8:00 a.m. insurgent attack in the Afghan village.

The village, the largest of Segall's sets, consists of two dozen ersatz mud-brick buildings, a small market, a rusting swing set, and, until recently, goats. (The goats were dismissed, because someone had to come in over the weekend to feed them, and more often than not it was Segall.) To get close to the action, I requested a role. I will be playing myself: a reporter who gets in the way and distracts people from their jobs. They've placed me in a sparsely furnished two-room house with a seasoned medical role-player named Caezar Garcia.

Under a torn pant leg, Caezar wears a simulated skin sleeve— silicone encrusted with mock gore and plaster bone fragments. A simulated severed artery will bleed via a small pump connected to three liters of house-brand special effects blood that Caezar wears in a concealed backpack, a sort of CamelBak for vampires. The flow is controlled by a wireless remote, so it can be stopped or slowed or allowed to continue unabated, depending on how competently the corpsman has placed the tourniquet. Originally the instructors, who hover on the fringes of the action during scenarios, held the remotes. Caezar, wanting a more nuanced bleed, petitioned to control it himself.

"I said, 'Look, once you bleed me out—'" Caezar stops to listen. The call to prayer has started. The recording, being played over a set of speakers on a tower at the center of the village, is the signal for the role-players and the pyrotechnics guy to take their places. Through a window to our left, the trainees can be seen entering the village. They walk in formation, armed and armored, looking unrelaxed. The tape-recorder muezzin finishes his call, and for a moment it's quiet. I can hear the soft, plasticky thrum of Caezar's blood pump.

And then I can't. First comes the familiar high whistle of an explosive-powered projectile, a sound that, depending on your life experience, presages pretty lights in the summer sky or a rocket-propelled grenade explosion. Rifle fire follows. The ammo is blanks, but you wouldn't necessarily know that, because the pyrotechnics guy sets off an accompanying "dust hit" on the ground or wall.

The muezzin's voice has been replaced by a recording of whizzing, ricocheting bullet noises and panicked soldiers yelling. It sounds like it was a hell of a battle. (I asked Segall about it later. "Vietnam?" "*Saving Private Ryan*.") You wonder what they make of it over at the Water Authority.

"OOOOOH, FUCK! AAAAAAOHH HELP ME!" That's Caezar. He's very good.

A trainee steps into the room. His gaze drops to the floor, to a foot, in a boot, nowhere near a leg. Bone and mangled flesh—the remnants of a lower leg, sculpted by "wound artists" working from photos of a real injury—protrude from the boot. The corpsman blurts out, "Are you okay?"

Years ago, crossing a street with my friend Clark, we looked down to see a smear of blood and feathers marginally recognizable as a pigeon. Clark bent over and yelled, "Are you okay?" The line is less funny now but equally ludicrous. A small blood lake expands on the floor. And here is where things go hyper-realistic: Unbeknownst to this corpsman, Caezar is an amputee.[*] He wears the silicone sleeve over the stump of his leg. When he jerks it around, as he is doing now, it trails an arc of blood. Blood is flying like champagne in the locker room after the big win.

Outside the door, instructors are yelling to get the other wounded

---

[*] And the founder of Missing Something, my second-favorite amputee organization name, after Stumps R Us. I attended a Stumps bowling party in the 1990s, which served as my official introduction to the awesomeness of Hosmer Upper Extremity Prosthetics sporting attachments. In addition to the Bowling Attachment, Hosmer makes a Baseball Glove Attachment, and the pole-gripping Ski Hand/Fishing Hand. The Hosmer-equipped bowlers kicked my ass.

"off the X"—out of sight, out of the kill zone. They're dragged into the room adjoining ours. The floor is men: role-players on their backs and trainee corpsmen crouched around them. One figure stands out for being unusually barrel-chested. This is the Cut Suit actor. You may be familiar with "patient simulators" like Resusci Anne, upon whom first responders practice their skills. The Strategic Operations Cut Suit is a "human-worn" patient simulator. The actor dons a vestlike rib cage with an insert tray of abdominal organs and, over this, a kind of flesh-tone wetsuit—simulated skin that bleeds when it's pierced, via the same pump-and-tube system Caezar uses for his stump. (It also "heals," with help from the Cut Suit Silicone Repair Kit.) It's as though someone crawled inside Resusci Anne and gave her the one thing patient simulators, for all their bells and whistles, will never have: humanity. SimMan may bleed and pee and convulse, his tongue may swell and his bowels may rumble, but he will never sit up, drill his gaze into a student's eyes, and plead, as Caezar just did, "Get me out of here, this is a *bad* neighborhood, man!"

Today's Cut Suit actor isn't yelling, because his character has been shot through the chest and his lung has collapsed. He takes shallow panicky breaths while a trainee, whose uniform identifies him as Baker, gets ready to do a needle decompression. When a bullet or broken rib punctures a lung, inhaled air begins to fill the cavity that houses the lung. The air builds up and soon the lung can't expand, and breathing becomes a struggle. It's called pneumothorax, from the Greek for *air* and *chest*, and it is the second most common cause of combat death. Baker's task is to insert a needle catheter to release the air and relieve the pressure. He's sweating. His glasses slide down his nose. He holds the needle near the role-player's collarbone, which is not between any of his ribs, or even part of the Cut Suit.

"Are you FUCKING SERIOUS, BAKER?" You know the exaggerated TV cliché of the scary yelling Marine instructor? It's not exaggerated. "That's his *clavicle*. You almost actually stabbed him."

Presently the needle finds its mark, an occlusive bandage is applied, and the role-player is loaded onto a stretcher. Baker picks up the stretcher's front handles without alerting the trainee at the other end, causing the patient and the $57,000 Cut Suit to tumble onto the ground.

"What the fuck is wrong with you, Baker?!"

Nothing, in fact. Just his sympathetic nervous system doing its job. Anything perceived as a threat trips the amygdala—the brain's hand-wringing sentry—to set in motion the biochemical cascade known as the fight-or-flight response. Bruce Siddle, who consults in this area and sits on the board of Strategic Operations, prefers the term "survival stress response." Whatever you wish to call it, here is a nice, concise summary, courtesy of Siddle: "You become fast, strong, and dumb." Our hardwired survival strategy evolved back when threats took the form of man-eating mammals, when hurling a rock superhumanly hard or climbing a tree superhumanly fast gave you the edge that might keep you alive. A burst of adrenaline prompts a cortisol dump to the bloodstream. The cortisol sends the lungs into overdrive to bring in more oxygen, and the heart rate doubles or triples to deliver it more swiftly. Meanwhile the liver spews glucose, more fuel for the feats at hand. To get the goods where the body assumes they're needed, blood vessels in the large muscles of the arms and legs dilate, while vessels serving lower-priority organs (the gut, for example, and the skin) constrict. The prefrontal cortex, a major blood guzzler, also gets rationed. Good-bye, reasoning and analysis. See you later, fine motor skills. None of that mattered much to early man. You don't need to weigh your options in the face of a snarling predator, and you don't have time. With the growing sophistication and miniaturization of medical equipment, however, it matters very much to a corpsman. Making things worse, the adrenaline that primes the muscles also enhances their nerve activity. It makes you tremble and shake. Add to this the motions and vibrations of a medevac flight, and you start to gain an appreciation for the military medic's challenges.

On top of caring for the wounded, corpsmen are expected to return fire if no one else is able. Like any precision task, marksmanship deteriorates in high-stress situations. The average police officer taking a qualifying test on a shooting range scores 85 to 92 percent, Siddle told me, but in actual firefights hits the target only 18 percent of the time.

The corpsman trainee working on Caezar is having difficulty with the tourniquet. Like Baker, he's a fine, fumbling example of the downside of an adrenaline rush. An instructor puts his head through the doorway. "What are we doing in here, fuckin' organ transplant? Let's go!"

If the scenario were real, Caezar would be dead by now. With a large artery bleed, it can take less than two minutes for the human heart—and, no coincidence, the Strategic Operations Blood Pumping System—to hemorrhage three liters: a fatal loss. The human body holds five liters of blood, but with three gone, electrolyte balance falls gravely out of whack, and there's not enough circulating oxygen to keep vital organs up and running. Hemorrhagic shock—"bleeding out"—is the most common cause of death in combat.

This is the grim calculus of emergency trauma care. The more devastating the wounds, the less time there is to stabilize the patient.

The less time there is and the graver the consequences, the more pressure medics are under—and the more likely they are to make mistakes. In a 2009 review of twenty-two studies on the effects of "stressful crises" in the operating room, surgeons' performance was reliably compromised: not only their technical skills but their ability to make good decisions and communicate effectively. And the stressful crises of the operating room—defined in this study as bleeding, equipment malfunctions, distractions, and time pressure—are business as usual in a theater of war.

Caezar exits the scene in a fireman carry, draped around a trainee's neck like a heavy mink stole. Baker follows behind with the stretcher. He's struggling because his palms are sweaty. He sets down his end in order to wipe his hands on his pants—again, without alerting the guy holding the other end.

"*Really*, Baker?" Palm sweat is a feature of fight-or-flight thought to have evolved to improve one's grip, but too much of it obviously has the opposite effect. "Put your fuckin' little girl gloves on if you have to."

The instructors are mean for a reason. They aim to subject the trainees to as much fear and stress as they can without actually shooting at them. The entire experience—the mock injuries, the gunfire and explosion sounds, the anguish of being called a little girl in front of everyone—is meant to function as a sort of emotional vaccine. Combat training for all troops, not just medics, has traditionally included exposure to some kind of simulated gore and mayhem. For years, writes Colonel Ricardo Love in his 2011 paper "Psychological Resilience: Preparing Our Soldiers for War," commanders have shown their charges photos and videos of gruesome injuries, or brought in veterans to talk about "the horrors they experienced." To help prepare future corpsmen, the Naval Health Research

Center hands out copies of *The Docs*, a 200-page comic book with lurid drawings of blast and gunshot injuries—a graphic graphic novel.

The pyrotechnics and battle soundtrack not only add realism but also kick-start the fight-or-flight reaction. Sudden loud noise triggers a cluster of split-second protective reflexes known as the startle pattern. You blink to protect your eyes, while your upper body swivels toward the sound to assess the threat. The arms bend and retract to the chest, the shoulders hunch, and the knees bend, all of which combine to make you a smaller, less noticeable target. Snapping the limbs in tight to the torso may also serve to protect your vital innards.* You are your own human shield. Siddle says hunching may have evolved to protect the neck: a holdover from caveman days. "A big cat stalking prey will jump the last twenty feet and come down on the back and shoulders and bite through the neck."

This may lead you to wonder, do impalas and zebras exhibit the startle pattern? And you would not be first in wondering. In 1938, psychologist Carney Landis spent some time at the Bronx Zoological Park, testing the evolutionary reach of the startle pattern, and the patience of zoo staff. In exhibit after exhibit, Landis could be seen setting up his movie camera and firing a .32-caliber revolver into the air. Less unsettling for zoo visitors—and more entertaining—would have been the experimental technique of fellow startle response researcher Joshua Rosett, who snuck up behind his (human) subjects and flicked the outer edge of their ear with his index finger. I imagine it was a trying time for the Rosett family.

The Bronx Zoo had no impalas, but they did have a goatlike

---

* But not your iPhone. Smart smartphone thieves use the startle reaction to their advantage. They come up behind unsuspecting texters and whap them on the back of the head. The startled victim's arms bend, launching the phone, which is effectively tossed to the thief.

Himalayan tahr, and it was duly startled. As was the two-toed sloth, the honey badger, the kinkajou, the dingo, the Tibetan bear, the jackal, and every other mammal that endured the scientific obnoxiousness of Carney Landis.

You will not be startled to learn that Landis's book-length treatment of the topic, *The Startle Pattern*, fell somewhat shy of runaway success.

TODAY'S SECOND scenario is a simulation of the aftermath of an explosion on a Navy destroyer. I have a symptom this time: smoke inhalation burns, which entitles me to some lines and a perioral dusting of soot. The set comprises a room of sailors' bunks, or "racks," and a sick bay down the hall. Catwalks overhead allow instructors to observe the trainees and hurl down invective.

The sight of smoke from a smoke machine is our cue to action. Five of us lie on racks in the dark, emoting amateurishly. I tell the trainee who comes to my aid that it hurts to breathe. He helps me out of my rack and steers me out to the hall. "Right this way, ma'am," he keeps saying, as though my table awaits. He shouts ahead that I'm going to be the priority. "Ma'am, we're going to have to crike you. Do you know that that means? We're going to make a small incision right here." He touches the front of my neck. Crike is short for cricothyrotomy. They're going to pretend to cut an emergency airway for me to breathe through.

"You are?" My symptoms only call for oxygen.

"Yes, we are. Because you can't breathe." I'm lifted onto the sick bay exam table.

"Well, it's more that it *hurts* to breathe." I'm trying to give a hint. "It *burns*."

The trainee picks up a scalpel. A voice sounds from above, like God calling to Abraham. "Stop!" It's one of the instructors. "She's talking to you, right? Then she's breathing. She doesn't need that."

Someone else yells, "Blood sweeps!" A corpsman trainee reaches under my back and slides both hands from shoulders to hips. He looks at his hands, checking for blood, for a wound that might have been overlooked. If you don't happen to be wounded, blood sweeps feel lovely.

My massage is short-lived. I'm carried back out to the hallway and set down beside another amputee actor, Megan Lockett. I saw Megan in the makeup room earlier. The special effects gore was still wet on her stump. She sat with her legs crossed, idly scrolling on her phone. It was like lions had come and gnawed off her foot while she checked Facebook.

The floor is slick with blood. Megan is having a bleeder malfunction. A pair of trainees skid and slip, trying not to drop the latest priority victim, a man wearing a tourniquet on his lower leg where a sock garter, in more civilized circumstances, might go. They plop him down on the exam table.

"And why is this guy so important?" yells God from on high.

"Open fracture!" someone tries.

"Is he dying? No, he's not!" More loudly now: "Who's dying, people? Who is the most likely to die?" No answer. God's hand points at Megan. Megan raises her stump. *Hello, boys!* "What does this patient look like she has?"

Two trainees rush over to get Megan, while Open Fracture joins me in the hallway of survivable maladies. I try to make some room, but my pants are sticking to the floor. I learn later that Karo syrup is the main ingredient in special effects blood. This makes life safer and more pleasant for actors whose role calls for them to cough up blood, but if it dries while you sit or stand in it, you will fuse to the floor like a candy apple on a baking tray.

When it's all over, the trainees are called to a debriefing on the pavement outside the set. An instructor named Cheech starts it off.

"That was godawful. You lost your minds. A woman who's missing a leg should have been the number one priority."

Excuses are offered. It was dark. Smoky. She was down on the floor.

"There was one patient standing in the middle of the room," Cheech says. "*Standing in the middle of the room.* And no one paid any attention to him. You need to make your bubble bigger. Don't get fuckin' tunnel vision."

The technical term for fuckin' tunnel vision is attentional narrowing. It's another prehistorically helpful but now potentially disastrous feature of the survival stress response. One focuses on the threat to the exclusion of almost everything else. Bruce Siddle tells a story about a doctor who had some fun with an anxious intern. He sent him across the emergency room to sew up a car crash victim's lacerations. The intern was so intent on his stitching that he failed to notice his patient was dead.

IT IS easy to get lost on the way to the Strategic Operations bathroom, and very entertaining. You might pass a rack of freshly painted excretory systems hanging in the sun to dry, or a man seated at a workbench, trimming the seams of a molded silicone Cut Suit penis.* You might overhear a person say to another person, "If you use different blood, it voids the warranty." At one point I take a wrong turn

---

* More formally known as the "optional integrated phallus," available in Caucasian and African American (different colors, same size).

and find myself in a storage area. A filing cabinet drawer is labeled "Spleens." "Aortas," another says. On the top of the cabinet, Cut Suit skins are folded like blankets. When I finally find the bathroom, the sign on the door, which uses the military slang "HEAD," confuses me in a way it would ordinarily not have.

Making my way back, I pass a Cut Suit training tutorial and decide to sit in. A woman with creamy tanned skin and variegated blonde hair stands at a table with the suit's various components, which she is demonstrating, like Tupperware, to two Marines from Camp Pendleton. (The Marine Corps had just purchased one of the suits, and the two Marines, Ali and Michelle, were training to be Cut Suit Operators.) The teacher, Jenny, shows them how to unsnap the "visceral lining" to access the abdominal organs. "You can do an evisceration," she says pleasantly, and notes that a slashed latex lining can be simply discarded and replaced.* Visceral Linings are available for purchase in packages of two hundred. It seems like a crazy amount of evisceration.

Jenny picks up a loose intestine and tells Ali and Michelle that they could, if they wished, fill it with simulated feces that they could make themselves, using oatmeal dyed brown and scented with a party novelty called Liquid Ass. The Cut Suit training coordinator, Jaime de la Parra, used to travel to conferences with Liquid Ass in his luggage, for demonstrations. Other employees, including Jenny, do not, and recently Jaime asked her why. "I told him: 'Because no one will come to our booth.'"

Segall, the Cut Suit's inventor, is proud of its realism, and justly

---

* Expendable items like Visceral Linings, Replacement Veins, Foreskins (for the Nasco Circumcision Trainer) and Laerdal's Concentrated Simulated Vomit are known in the industry as "consumables." In the case of the Simulated Boluses of chewed food that get stuck in the esophagus of the Laerdal Choking Charlie manikin, the term is doubly apt.

so. Still, no matter how rank the intestines smell or how realistically the amputee's stump is bleeding, students must know it's not real. No one hacks off a limb to train a group of medics.

Or not a human limb, anyway.

A S FAR back as the 1960s, students of combat trauma medicine have practiced life-saving procedures on anesthetized pigs and goats. There would be no issue here, except for the fact that barnyard animals don't naturally wind up in situations where they're shot or stabbed or blown up by an IED. So the only way to train students on them is to hire a company to do the shooting or stabbing or leg-removing. There's one of those companies not far from here.

Live tissue training is the topic of conversation at lunch today, on the back deck of Stu Segall's diner. Stu and I are joined by Kit Lavell, the company's executive vice president. Lavell fills me in on legislation that would require the Department of Defense to reduce the number of animals used for live tissue training from the 2015 level—about eighty-five hundred per year—to somewhere between three and five thousand. An animal rights organization called Physicians Committee for Responsible Medicine is behind the push. Advances in patient simulators—and high-drama Cut Suit demos before members of Congress—have made it harder for defenders of live tissue training to make their case.

Unfortunately for pigs, the layout and size of their viscera approximate ours, as do their blood pressure and the rate at which they bleed. Goats are better for practicing emergency airway procedures, as there's not four inches of neck fat to slice through.

I watched a YouTube clip purporting to be part of a live tissue training class that someone surreptitiously filmed. A group of men stand around a folding table on a rainy day. A makeshift roof with a tarp drips overhead. Two or three men at a time lean over an inert pig laid out on the table. Their backs are to the camera. They chat quietly. They look like pit masters at a whole-hog barbecue. A veterinarian is there, and you can hear someone ask him to give the animal a bump, meaning more anesthesia. The leg amputation happens off-camera, but you can see the instrument the instructor uses: a set of long-handled shears of the sort one might use to cut through chain link. It sounds ghastly but gets the job done quickly. Assuming the anesthetic was competently administered, the proceedings struck me as no more upsetting than what goes on in slaughterhouses every day in the name of bacon and chops and short rib ragu.

For that very reason, Siddle feels, it's an incomplete "stress inoculation." "While it's a good experience to work on something live, something that pumps, it's not a human. It's not screaming." To gain experience with actual screaming humans, Camp Pendleton's corpsman trainees may spend time observing and helping out in an emergency room in a gang-saturated Los Angeles neighborhood. "That's our equivalent of Iraq or Afghanistan," Ali said earlier. "Gunshots, strafings, stabbings."

Michelle, the other Cut Suit Operator-in training, experienced both live tissue training and a stint in an emergency room. She found them helpful in different ways. Live tissue training provides a controlled teaching environment. Students can try things out, grab a slippery artery between two fingers to stanch a bleed. "You're not," she said, "going to be doing that with a patient in an emergency room."

With its bleeding, wheezing, cursing role-players, Strategic Operations tries to be one-stop shopping: something pumping, human, *and* screaming. "It creates a willful suspension of disbelief," says Stu, disarticulating a fried fish. I don't quite understand that phrase, but I do understand what he says next. "We've had students wet themselves, soil themselves, vomit, faint."

Lavell shares that Dennis Kucinich lost his congressional lunch at a Cut Suit demo. The representative from Ohio was sitting in the front row with his wife, Elizabeth, the prominent DC vegan and animal rights advocate. "When the actor started screaming and the blood started spurting, Kucinich went white. You could see the reverse peristalsis beginning." I glance at neighboring tables, half expecting to see some here. "His wife got up and helped him to the door."

T**HE MAIN** stressor of combat medicine is absent from every training simulation. No one is shooting real bullets at or anywhere near you. "Training is limited by liability," said Siddle. He sounded a little mournful.

"The high number of returnees diagnosed with PTSD suggests we are not doing enough," scolds Colonel Ricardo Love in his paper. Love hailed the ancient Spartans' approach to "building psychological resilience in their forces." *Pelopidamus, looketh upon these novel strategies for building resilience.* "On several occasions [the] war games were deadly and some boys were killed." According to Sparta scholar Paul Cartledge, other military resilience-builders included the stalking and killing of random slaves and "the braving of whip-lashing seniors* in order to steal the largest possible number of cheeses from the altar of (Artemis) Ortheia, a goddess of vegetation and fertility."

---

* I suppose that by "seniors" Cartledge means people older than the boys; however, Spartan senior citizens weren't the courtly walker-pushers of current stereotype. "Tribal elders" would screen babies for military worth; those deemed unfit were hurled into a chasm called "the deposits." Nothing in antiquity makes much sense. Who gives cheese to a goddess of vegetation?

Many years ago, reporting a story on killer bees, I experienced a kind of stress inoculation. I accompanied a team called out to remove a hive on a farmer's land in south Texas. The venom of "killer" honeybees is the same as that of ordinary honeybees, but the bees are far more aggressive in their defense of the hive and their pursuit of interlopers. The larger the hive, the more defensive the bees. This hive filled a fifty-five-gallon oil drum. I wore a bee suit, but I hadn't attached the veil properly and bees began getting underneath it and stinging me. Later that day I and my throbbing welts visited a keeper of ordinary honeybees. While we talked, bees would light on my arm. My normal reaction would have entailed flailing and girly alarm noises. Instead I calmly watched them crawl around. Fear of bees: gone.

But would it have worked in reverse? Would exposure to regular honeybees have inoculated me against the fear I felt inside the killer bee swarm? Caezar's theatrics and Tom Hanks yelling and the hectoring instructors—these are regular honeybees. Still, as Siddle allows, "Anything that narrows the gap is good."

The other way to train medics is to have them practice a skill so many times that it becomes automatic. So when the prefrontal cortex goes AWOL, when reasoning drops away, muscle memory, one hopes, will persist. Do it enough times, and you can administer first aid in the ultimate survival stress scenario: when the gore is your own. Recall the combat engineer from chapter 4 who'd stepped on an IED. "Without thinking"—as he aptly put it—he pulled out a tourniquet and placed it perfectly on what remained of one leg.

CAN THE carnage of an explosion ever really *not* be stressful? Does a disarticulated head ever come to seem normal? Apparently. "After a while," Ali told me during a break from the tutorial, "it's just a head. You get on with your job." Michelle told a story from her deployment in Iraq. She was carrying part of a Marine's leg that had been blown off by an IED. The foot was still in the man's boot, and presently his buddy went to pull it out. When the boot relinquished its hold, the foot smacked Michelle in the face. She made a face that led me to assume the foot had started to decompose. "It wasn't decomposed," she said. "It was a brand-new, blown-off foot." She leaned closer. "He wasn't wearing socks." What repelled Michelle was not blood or gore, not the foot's detachment from the rest of the body or the awful deadness of it, but the smell and feel of the sweat on her cheek.

And that will serve as my lurching segue to the miraculous, reviled excretions of the human eccrine gland. In a place like Afghanistan, sweat keeps more people alive than corpsmen do.

7 Sweating Bullets

The war on heat

FORT BENNING, GEORGIA, HAS three key ingredients for heatstroke: humidity, intense sun, and Army Ranger School. Rangers, like their better-known cousins Navy SEALs, are part of the US Special Operations forces. To borrow the words of their creed, the Ranger is an "elite soldier" expected to "move further, faster and fight harder than any other soldier." Josh Purvis would seem to be maximally elite in that he was, when I met him, an instructor at Army Ranger School and a contender for the annual Best Ranger Competition. The competition falls into the category of a multisport event, surely the only one to include a Bayonet Assault Course and a litter carry. (They don't mean trash.) Competitors march and run twenty-plus miles with a sixty-pound pack, and every year, a few will experience a second litter carry, in the horizontal position. In 100-degree heat, "further, faster" can be a lethal undertaking.

Today Purvis, along with a fellow instructor, will march in a hot spell of mechanical making. As subjects in a heat tolerance study, they will walk fast and uphill for two hours at 104 degrees

Fahrenheit on a treadmill inside the "cook box" at CHAMP: the Consortium for Health and Military Performance, part of the Uniformed Services University of the Health Sciences. Some individuals are constitutionally prone to heatstroke and other heat illness, and the Army would like to have a way to know who they are before sending them out into the scorch of a Middle East afternoon with a hundred pounds of gear and other human beings whose lives depend on them.

Purvis leans against a counter, shirtless, filling out a "mood state" questionnaire. I see him put a check in the "Moderately" column beside the descriptor "full of pep." *Pep* isn't really a word for Josh Purvis. *Pep* wants some spring to the step, some twinkle, a tendency to whistle. I don't believe Josh Purvis whistles. His features, while handsome, have a hard set to them, a sort of coiled restiveness.

The researcher, a fit blonde with a luminous complexion, tells Josh to make a fist. She and her colleagues are looking for biological and genetic markers that might lead to a simple blood test to identify combatants prone to heat illness, so their superiors could keep closer tabs on them. However, her request is unrelated to the drawing blood. "Josh, let us see your muscle." The researcher is Josh's mother, Dianna Purvis.

Josh's arms remain at his side. "*Mom.*"

Purvis the elder holds out an apple from the pre-test meal pack. "Josh, eat your snack before you go in."

"Mom, *stop.*"

I can't see which box Josh has checked beside the descriptors Uneasy, Peeved, and On Edge, but I'd mark them "a little bit." His mother has put this down to "the probe." He will be having his tolerance tested—heat and otherwise—by way of a rectal probe: a slim, flexible, insertable thermometer. The rectal probe is attached by a six-foot wire to a piece of portable hardware labeled Physitemp Thermes. It is the size of a hardback book, and heavy as a brick. It's heavy enough that if you set it down on a counter, forget that you are tethered to it, and walk away, you will be very *effectively* halted before you drag it off the counter.

The rectal thermometer enables the researchers to monitor their subjects' core temperature. Like any complex bioelectrochemical system, the human body works best when its vital components are humming along in a set temperature range. For humans, that's roughly ninety-seven and a half to ninety-nine and a half degrees Fahrenheit. When your core temperature begins to rise, either because it's hot where you are or you're toiling hard, or both, the body takes measures to bring it back to the happy range. First and foremost, it sweats.

Until this trip, I thought of sweat as a sort of self-generated dip in the lake. But sweat isn't cool. It's warm as blood. It essentially *is* blood. Sweat comes from plasma, the watery, colorless portion of blood. (A dip in the lake cools by conduction: contact with something colder. Highly effective but not always practical.) Sweat cools by evaporation: offloading your heat into the air. Like this: When you start to overheat, vessels in your skin dilate, encouraging blood to migrate there. From the capillaries of the skin, the hot plasma is offloaded through sweat glands—2.4 million or so—onto the surface of the body to evaporate. Evaporation carries heat away from the body, in the form of water vapor.

It is an efficient system. A human in extreme heat can sweat as much as two kilograms an hour, over a span of a few hours. "Roughly speaking, 10 kilograms loss of sweat [over the course of a day] is not rare for workers in overheated factories and active soldiers stationed in the tropics," states the late Yas Kuno, longtime professor of physiology at Nagoya University School of Medicine, in the 1956 edition of *Human Perspiration*. "One will be struck with wonder . . . when he thinks that such a large amount of sweat is produced from glands which are extremely small in size." Though humans have, by weight, more than twice as much salivary gland tissue as sweat gland tissue, they are capable of producing six times as much sweat as spit.

Human Perspiration is itself a prodigious output: 417 pages. There was a lot to report,* in part because Kuno's sweat studies spanned thirty years, and in part because he had a lot of collaborators: "some 65 in all." The book includes a collection of black-and-white photographs of Japanese men in thongs, sweating after a session in the Perspiration Chamber. Because the men had been dusted with a special starch that turns black on contact with sweat, their torsos, foreheads, and upper lips are speckled with what appears to be an especially virulent mildew. One set of images highlights the surprising variety of sweat distribution patterns on the human scalp.† Rather than take a razor to their own heads, the researchers recruited "eighteen

---

* Kuno and his team spent a great deal of time exploring the differences between thermal and emotional sweating, the latter wetting the palms and soles and the former, everything but. One researcher excised a patch of leg skin and grafted it to his palm. Would the patch henceforth, unlike normal leg real estate, sweat when the man was nervous? (Yes.) Would it remain dry in emotionally trying circumstances, such as when colleagues tittered over the sudden and suggestive appearance of hair on one's palms? (No.) The emotional sweat work conferred a corollary talent for lab-based sadism. The researchers invented and delivered terrible news to their subjects. They tasked them with oral arithmetic problems. They threatened to administer painful shocks, provoking "the uneasiness of expecting pain." Kuno was the Stanley Milgram of perspiration.

† The human head sweats like a mother. As the cradle of the brain, it's served by a lot of blood vessels, and those vessels, unlike the vessels of the body's other extremities, don't constrict. Thus head wounds bleed readily and faces flush and sweat. But it's misleading to say, as one so often hears, that people lose 90 percent of their body heat via their head. "My father-in-law, when he sees me go out in winter with no hat, always tells me that," says military research physiologist Sam Cheuvront. "I say, 'If that's true then I should be able to put on a tassel cap and go outside naked and retain 90 percent of my body heat.'" When in fact, he'd be losing heat through his exposed body parts. Though gaining my affection.

Buddhist Japanese priests who make it a custom to shave their heads" and, going forward, to ignore all calls from Nagoya University.

Outside of thermoregulation labs, sweat commands little respect, a fact that needled Kuno. "It is peculiar," he wrote, "that the value of sweating is appreciated only by patients [who cannot sweat], who suffer greatly from heat, and not by ordinary people, who usually complain about too much sweat." *Jerks.* To Kuno's mind, nothing less than the march of civilization was forged by the indomitable human thermoregulatory system. "The human race inhabits the whole earth, . . . while the living zones for most animals are more or less confined. This privilege of the human race has partly been acquired by their intelligence, but their spreading over the torrid zone has only succeeded through the high development of the sweat glands." Were it not for human perspiration, there would have been no Vietnam War, no Operation Iraqi Freedom, no Georgia-based Army Ranger School.

If sweating is so effective, why were there 14,577 cases of heat illness among active US Armed Forces personnel between 2007 and 2011? Because they work too damn hard. When sweaters exert themselves, the muscles they're using begin to demand the blood that the body needs to use for sweating. The mildest consequences of this competition for blood are heat exhaustion and heat syncope—fainting. With blood flowing out to the skin for cooling purposes and, at the same time, into the muscles to deliver oxygen to fuel the body's toil, it becomes harder to maintain the blood pressure needed to pump blood up to the brain. Without enough oxygen-carrying blood reaching your brain, you pass out. (Counterintuitively, overheated people sometimes pass out not in the midst of their exertions but when they stop and stand still; this is because contracting the leg muscles helps keep blood from pooling down there.)

Heat exhaustion is embarrassing but not particularly dangerous.

Fainting is both symptom and cure. Once you're horizontal on the ground, the blood flows back into your head and you come to. Someone brings you water and escorts you to the shade and you're fine.

Heatstroke, however, can kill. Here too, it begins with a competition for blood. On a hot day, when your body is trying to sweat your core temperature down to the safe range and you haven't been drinking enough water to replenish your blood volume, and on top of that you're exercising hard and your muscles are clamoring for oxygen—and the exercise itself is generating heat—something has to give. "The body sacrifices flow to the gut in order to put it where it's needed," explains Sam Cheuvront, a research physiologist at the US Army Research Institute of Environmental Medicine (USARIEM), part of the Natick Labs complex. The splanchnic organs—a stupendously ugly way to say *viscera*—are cut off from the things they need: oxygen, glucose, toxic waste pickup. The technical term is ischemia. It is a killer. The digestive organs start to fail. The gasping gut may begin to leak bacteria into the blood. A systemic inflammatory response sets in, and multi-organ damage ensues. Delirium, sometimes coma, even death, may follow.

Other scientists emphasize heat damage to the central nervous system: Brain proteins unfolding—"denaturing" is the technical term—and malfunctioning. (When you cook an egg or a piece of meat, the change in texture is caused by denaturing.) Cheuvront doesn't buy the "hot brain" theory. Protein denaturing, he said, occurs at temperatures much higher than the 104 degrees Fahrenheit of a heatstroked brain. There are hot tubs in Japan hotter than that. Cheuvront indicated that there's no real consensus on how heatstroke kills. Except this: "Lots of bad things happen."

Gut ischemia may help explain why the US military life raft survival food packet appears, at first glance, to be a cruel joke: nothing to eat but packages of colorful old-timey sour balls, brand-named Charms.* If you're baking on tropical seas and your digestive organs are shutting down, you are not impelled to eat. One thing to be said for sour balls: The acidity stimulates saliva flow, a welcome feature for dehydrated, cotton-mouthed lifeboaters.

HUMIDITY IN the cook box is set at a highly bearable 40 percent, which goes a long way toward explaining why I'm still vertical. When the air around you is saturated with moisture, your sweat—most of it, anyway—has nowhere to evaporate to. It beads on your skin and rolls down your face and back. More to the point, it doesn't cool you. In the 1950s, the US military invented an index for the treacherousness and downright god-awfulness of heat, called wet-bulb globe temperature: wind chill factor's partner in meteorological misery. WBGT reflects the varying contributions of air temperature, wind, sun strength, and humidity. Humidity is a full 70 percent of it.

It's the humidity, but it's also the heat. When the air is cooler than 92 degrees Fahrenheit, the body can cool itself by radiating heat into the cooler air. Over 92—no go. Radiation's partner is convection: That cloud of damp, heated air your body has generated rises away from your skin, allowing cooler air to take its place. And, provided it's drier, allowing more sweat to evaporate. Likewise, a breeze cools you by blowing away the penumbra of swampy air created by your

---

* Charms used to be part of ground rations, too. They were removed partly because of a persistent belief that they brought bad luck. No one at the Natick Labs Combat Feeding Directorate knows the origins of the unlucky-Charms superstition. I like this guess best, from the gun-enthusiast website AR15.com: "Because the plastic wrapper sticks . . . and results in you getting drilled in the brainpan because you were picking at a piece of candy and not paying attention."

body. If the air that moves in to take its place is cooler and drier, so, then, are you.

After fourteen minutes in the cook box, I'm sweating lightly. Josh Purvis, on a treadmill behind me, began sweating much sooner than me. The hair on his forearms is matted to his skin. The dragon on his chest appears to be weeping. I took all this to be an indicator of his inferior heat tolerance, but in fact the opposite is true. People who are heat-acclimated typically, as Dianna Purvis puts it, "sweat early and copiously." Their thermoregulatory system takes action swiftly. Mine took ten minutes just to figure out what was happening. *Hey, is it hot in here? Should I be doing something? I would enjoy a Popsicle right now.*

Not only is Josh better acclimated to the heat and humidity, he's vastly fitter than I am. Aerobic fitness and percentage of body fat are thus far the only factors shown to reliably set people apart in terms of their tolerance for heat. A strong heart pumps more blood per beat, making it more efficient at delivering oxygen to the muscles. That leaves more blood for the rest of the body and for making sweat. This doesn't mean that fit individuals don't get heatstroke. In the military, it's often the fittest who fall prey to exertional heatstroke, because they're the only ones capable of pushing themselves hard enough to reach that point.

"Are you ready for the pack?" Dianna has put thirty pounds of sandbags in a backpack to give me a sense of the weight that a soldier in Afghanistan would carry on a two-day ruck march. The typical combat load has been more than twice that—95 pounds, including 33 pounds of body armor, 16 pounds of batteries, and 15 pounds of weapons and ammunition. World War II–era desert survival experiments by Edward Adolph determined that carrying a pack half that weight caused a man to sweat an additional half pound of fluids per hour.

My pack holds only sand. I wear no body armor and carry no weapon other than a Thermes rectal probe. I don't know what mission this qualifies me for, but whatever it is, I'm in no shape or mood to undertake it. Within seconds of donning the pack, my heart rate shot up by 25 beats per minute. "You've increased your workload, so now you need a lot more blood to your working muscles." Dianna is yelling over the sound of the fans. "And your core temp is climbing. You're at thirty-seven point nine." 100.2 degrees Fahrenheit. Marching toward collapse.

Exacerbating the scenario is my tendency to underhydrate. I am what's known in the parlance of the Heat Research Group of USARIEM as a "reluctant drinker." Allowed unlimited access to water, a reluctant drinker in a perspiration chamber will quickly lose more than 2 percent of her body weight. And you can't trust thirst to tell you how much water you need to be drinking. Yas Kuno cites studies in which men hiked for three to eight hours without water, after which they were allowed as much water as they wanted. They tended to stop as soon as their thirst felt quenched. On average, that happened after drinking about one-fifth the amount of fluid they had sweated away.

Outside the cook box a big blue plastic tub is filled with cold water for anyone whose temperature passes 103. Immersion in cold water is the quickest fix for heat illness. When a hot solid or liquid comes in contact with a cooler one, the hot one will become cooler and the cool one hotter. That's conduction. Conduction explains why tropical shipwreck survivors can die of "warm water hypothermia." As long as the sea is cooler than they are, they lose body heat to the water.

Conduction can also, of course, make a body hotter. Should you find yourself stranded in the desert, don't rest directly on the ground—or lean against the Land Rover. Sand gets as hot as 130 degrees Fahrenheit; metal well hotter. Conduction helps explain why loose clothing keeps you cooler in the sun. A baggy shirt heats up, but because the cloth is not in contact with your skin, it does not—unlike a form-fitting t-shirt—conduct that heat to your body. (Loose clothes also let sweat evaporate more readily.)

Even better if the baggy shirt is white. Light-colored clothing reflects some of the sun's radiation, so you get hit with less of it. Going shirtless in the sun makes a person hotter, not cooler. In Edward Adolph's "'nude' men in the sun" study, subjects sitting on boxes wearing nothing but shoes, socks, and underwear suffered the equivalent of a ten-degree rise in air temperature. Adding to their discomfort, the control—a fully clothed man—was seated beside them. It's not the heat, it's the humiliation.

You can imagine how heat illness experts feel about sunbathers: people who willingly lie in direct sun, on hot sand, nearly nude. Small wonder it's done in close proximity to the big blue tub known as the ocean. Just don't get up off your towel and start lifting weights. Overworking a set of muscles puts you at risk for a potentially fatal condition called rhabdomyolysis. If the body can't keep pace with a muscle's extreme demand for fuel, eventually the muscle becomes ischemic. Heat exacerbates the scenario, because of the competing demand for plasma for making sweat. The cells of the oxygen-starved muscle tissue begin to break down, and their contents spill into the bloodstream. One of these breakdown components is potassium; high levels of it can cause cardiac arrest. Another, myoglobin, damages the kidneys—sometimes to the point of failure. Now you are a very buff and picturesque corpse.

Bodybuilding has been the number one pastime on bases in Afghanistan, where it is even hotter than in Venice Beach. The bodybuilding supplements soldiers take to bulk up more quickly exacerbate the risks. They often contain potentially dangerous compounds: stimulants that spur muscle contractility, thermogenic agents that rev the metabolism, and creatine, which accelerates dehydration. All of these increase the competition for the body's limited blood supply. CHAMP runs an online resource, Operation Supplement Safety, that reviews the dangers of different products; however, with more than ninety thousand different supplements for sale on the Internet—and Amazon.com delivering to the major air bases—it's a Sisyphean challenge. For those unfamiliar with the myth, Sisyphus was that Greek guy the gods punished by condemning him to roll an enormous boulder uphill forever, or until rhabdomyolysis set in. During 2011, there were 435 cases of exertional rhabdomyolysis among US service members.

Even simple protein supplements amplify the risks. Protein is deliquescent: It draws water from the body's tissues into the bloodstream to help flush the protein breakdown products, which are tough on the excretory system. If you're dying of thirst in the desert, drinking your urine won't help you. The proteins and salts are by that point so concentrated that the body needs to pull fluid from the tissues to dilute them, which puts you back where you began, only worse, because now you are saddled with the memory of drinking your own murky, stinking pee.

Rhabdomyolysis also turns up at the other extreme of the bodybuild spectrum. Morbidly obese patients immobilized on their backs—say, for lengthy gastric bypass surgery—run the risk that their bodies will press down on the muscles of their backsides so hard that circulation is cut off. After four to six hours, the dying cells of the muscle tissue break open and leak, and when the patient finally moves, or is moved, the blood rushes back in and sweeps the breakdown products into the bloodstream in a sudden, over-whelming gush. Being pinned under rubble in an earthquake or in the wreckage of a car poses a similar risk. As does passing out drunk and lying without moving for six hours. This was explained to me by rhabdomyolysis researcher Darren Malinoski, an assistant chief of surgery at the Portland VA Medical Center. He added that rhabdo-myolysis is one reason people roll over in their sleep. "The muscles are getting ischemic, and they make you move."

"Look: Even your thighs are starting to flush," says Dianna. All that overheated blood being shunted to my skin. "Do you want to try to keep going a full half hour with the pack on?"

Not even slightly. "I think I get it."

Dianna asks the lads how they're feeling. Josh's fellow instruc-tor, whose name is Dan Lessard, replies that he's bored. Josh doesn't hear the question because he's got earbuds in. He removes one, and a tinny musical aggression leaks out. It's Five Finger Death Punch, a metal band that from what I can tell uses synthesized machine-gun fire in place of a drummer.

Josh says he and Dan plan to do "a real workout" later in the day.

"Mary stopped after seven minutes with the pack on," Dianna volunteers. *Hey!*

Josh defends me. "You don't come out of the womb with a ruck-sack on. The first time I put it on, I hated my life." He seems like a good person who has been handed a lot. His frivolity, his pep, what-ever innocence we're all born with, became something tougher in Iraq. War denatures people.

At 11:30, we're released from the cook box. "And now you can go take out your friend," says a lab tech named Kaitlin, referring to the probe. Earlier, in the midst of a conversation about idiosyncratic sweating patterns, Kaitlin raised both arms as though she'd just won

Wimbledon and announced, "My right armpit sweats way more." This we confirmed. Which bring us to the point of Dianna's work: Genetic differences in thermoregulation—efficient/inefficient, left side/right side, you name it—are surprisingly large and well worth paying attention to, given our seemingly permanent posture of fighting extremism in the Middle East.

Dianna suggests heading to a nearby Walter Reed cafeteria to continue the conversation. Josh seconds. "*Sustenance.* Let's get it."

THE PIZZA at Warrior Café does not look healthy. By that I don't mean that it's unhealthy to eat it—though it possibly is—but rather that the item itself looks in poor health. The edemic crust. The sweating cheese. The scabs of pepperoni. I follow Josh and Dan to the salad bar. Like many in the US military, they are disciples of CrossFit, a workout that emphasizes real-world, or "functional," strength over isolated muscle development. And lots of garden greens.

"Everybody wants to get big and look strong," Josh says between mouthfuls, after we're seated. He eats with purposeful intensity, the way he speaks or strides on a treadmill. By "everyone" he means today's infantry. "There are different ways to do that. You can work hard, or you can do the bodybuilding thing, because you don't care about anything other than looking good. Nobody wants to work. They experiment with steroids. They want to be bigger, faster." The eyes fixed on the salad. "But that's not functional strength. And they have to lug it around, that muscle, and they have to cool it . . ."

"And the supplements themselves increase the risk of heat illness," I hear myself saying.

That's not Josh's concern. His concern is this: Unfit soldiers put the rest of the unit at risk. He places it in context for me: a hypothetical mission to clear and secure an insurgents' compound. "How about this. In the middle of a firefight, where you're already physically sucking, one of your buddies gets shot. You've got a casualty collection point in the first room that you cleared, but to get there, you have to drag him in his body armor. You're already smoked, and now you're dragging dead body weight, so now you're really smoked." He jabs at salad. Lunch is a syncopation of hunger and spite. Stab, shovel, chew, speak, stab. "Are you ready to deliver some first aid to this guy who's depending on you to save his life after you just got your ass handed to you, because you wanted to go do some curls at the gym?"

There is quiet at the table. I'm thinking this story maybe isn't hypothetical. I'm adjusting to the concept of a "casualty collection point," to the horrible fact that there can be enough casualties for a "collection."

"So," Dianna says after a moment. "Back to heat."

"I'm sorry." Stab, *stab*, shovel, chew. "I have very little to say about heat. People used to ask me, 'What it's like in Iraq?'" A garbanzo bean dies on a tine. "Open your oven and crawl in."

Dianna persists. "So Josh, I hear stories of guys superhydrating ahead of time so they don't have to carry water. So they can carry extra ammo."

Dr. Adolph looked into this. "By predrinking," he wrote, "man converts his interior into an accessory storage tank. A man on foot can thus carry as much as a quart or more of additional water." Adolph had a group of men fill their tanks by drinking two pints of water, and then sent them out into the heat on a "dehydration hike." By checking the dilution of their urine, he was able to conclude that only 15 to 25 percent of the "predrunk" water had been peed out. The rest was available to become cooling sweat.

However, desert survival scenarios aside, the US Army does not advocate predrinking. Exerting oneself on a sloshing stomach is uncomfortable and compromises performance. And soldiers who get carried away in their effort to fill the "storage tank" risk water poisoning: overdiluting the body's salt levels and throwing the system fatally out of whack.

Also, it may not be manly. "If you take extra ammo," says Josh in response to his mother's query, "you don't take it at the expense of water. You just take it. You man up, and you take it." Somewhere Josh found blueberries for his salad. He goes at them with brisk, well-centered stabs. He's going to ace the Bayonet Assault Course.

Speaking of hot-weather military dilemmas: Let's talk about body armor. The current ensemble weighs 33 pounds. You are weightlifting just going up a flight of stairs. Josh had a buddy who was killed on a rooftop without his body armor on. "His command was ridiculed for it. But in all reality, I wouldn't have had my body armor on, either."

"Do you take it off because it's too hot?" I'm like a fly buzzing around his head. A little yapping dog at his ankles.

"I take it off because it makes sense."

Dan steps in to lighten the tone. "Mary, we're walking up and down mountains with a hundred pounds on our backs, fighting guys in sandals and man-dresses. The Army's answer to a lot of things is to give you more equipment, more stuff, most of which takes batteries, and there's only so much you can carry."

The Army's other answer, one they have flirted with for over a decade now, is a wearable hydraulic exoskeleton to help with heavy loads. Lockheed Martin posted a video of its entry, the HULC (Human Universal Load Carrier), on YouTube. Soldiers are shown bounding across gullies and taking cover behind boulders while wearing articulated metal braces on the outsides of their legs, as though the Army had taken to conscripting 1950s polio victims. The HULC was tested at Natick in 2010, on a "prolonged march" with an 87-pound load. One of the comments posted for the You-Tube video comes from a participant in that test: "Everyone was pretty much done at forty-five minutes due to shin splints. " Others questioned whether fighters could move quickly under fire or even pick themselves up if they fell. Patrick Tucker, the technology editor for the website Defense One, tripped over the battery life: five hours, provided you're moving slowly (2.5 mph) and on level terrain. He doubted HULC's usefulness in places without a steady power supply—"like, basically any place where soldiers might, you know, have to fight."

"Do you want to know why my friend got killed?" says Josh. "Somebody probably heard him going into the building, because he couldn't be quiet enough, because he's carrying too much shit in the first place. There's all kinds of restrictions that risk-averse people are making. They have good intentions but they have bad effects."

Dianna points to my tape recorder. "You can probably turn that off." Heat isn't going to be a topic, just a mood.

DRIVING BACK from lunch, Josh and Dan sit in the back, planning their workout. I hear Dan say, "one hundred snatches," which hits my ear like a Dr. Seuss title. Up in front, Dianna and I talk science. I tell her about my recent visit to Natick Labs. *They have a manikin that sweats!* And "water-needs prediction equations"! You plug in the weather report and the fighters' loads and activity levels, and it tells you how much water you're going to need to haul to the battlefield. How excellent is that, I want to say, but I know Josh is listening.

I understand his scorn. I understand there's always a factor left out of the equation, something unknowable to someone who's not out there, inside the madness. I know every mission has unique requirements and risks. I know why there are derogatory names for people who sit in air-conditioned offices making rules for people out humping artillery across an open courtyard at noon in Afghanistan. Though at the moment, I can't remember what those names are.

"Chairborne Ranger?" offers Dan. "Pogue?"

"*Scientist*," says Josh. Dianna taps the steering wheel with one thumb. She glances in the rearview mirror. "I love you, son."

Josh stares out the window. "I love you too, Ma. No shame."

A few words in defense of military scientists. I agree that squad leaders are in the best position to know what and how much their men and women need to bring on a given mission. But you want those squad leaders to be armed with knowledge, and not all knowledge comes from experience. Sometimes it comes from a pogue at USUHS who's been investigating the specific and potentially deadly consequences of a bodybuilding supplement. Or an army physiologist who puts men adrift in life rafts off the dock at a Florida air base and discovers that wetting your uniform cools you enough to conserve 74 percent more of your body fluids per hour. Or the Navy researcher who comes up with a way to speed the recovery time from travelers' diarrhea. These things matter when it's 115 degrees and you're trying to keep your troops from dehydrating to the point of collapse. There's no glory in the work. No one wins a medal. And maybe someone should.

8 Leaky SEALs

Diarrhea as a threat to national security

SHOULD YOU ONE DAY travel to the overlooked desert nation of Djibouti, you will see from the window, as you land, what appears to be a large construction site adjacent to the airport. In fact, it's a US military base, Camp Lemonnier: 3,500 people who live and work in retrofitted shipping containers, some stacked, some side by side, a Tetris of unadorned rectangular boxes. Other than the shrubs that grow in the drip from the air-conditioning units, there is no landscaping. Interior décor takes the form of emergency instruction placards ("Stop and listen to the Giant Voice . . .") and framed chain-of-command portraits. In three days on base, I've seen a single item that one might class as luxury: one indulgent, cushy, costly item shipped here for no other reason than to add a little comfort to a soldier or sailor or airman's life. Captain Mark Riddle requisitions Charmin Ultra Soft for the container that belongs to Naval Medical Research Unit 3. The sign on the door explains it: Diarrhea Clinical Trial.

The word alone makes people want to laugh: *diarrhea*. Riddle doesn't fight this. On the contrary. He recruits study subjects through GOT DIARRHEA? signs on the backs of restroom stall doors. One of the photographs on the Stool Grading Visual Aid he created for participants in the current study comes from a Campbell's Chunky soup ad. ("Look closely," he'll confide, "there's a spoon sticking out.") Nevertheless, for reasons you will come to understand, Riddle takes diarrhea very seriously. As he has put it, intending nothing funny, "I live and breathe this stuff." I have heard him use the word *sacred* to describe a collection of frozen stool samples. Riddle would like military brass to take it seriously, too.

In past centuries, this took no convincing. Dysentery "has been more fatal to soldiers than powder and shot," wrote William "Father of Modern Medicine" Osler in 1892. ("Dysentery" is an umbrella term for infections in which the pathogens invade the lining of the intestine, causing cells and capillaries to ooze their contents and creating dysentery's hallmark symptom, the one that sounds like British profanity: bloody diarrhea.) For every American killed by battle injuries during the Mexican War of 1848, seven died of disease, mostly diarrheal. During the American Civil War, 95,000 soldiers died from diarrhea or dysentery. During the Vietnam War, hospital admissions for diarrheal diseases outnumbered those for malaria by nearly four to one.

Once germ theory gained acceptance and the mechanics of infection became known, microorganisms—and the filth they breed in, and the insects that deliver them—became targets of military campaigns. Suddenly there were Fly Control Units, sanitation officers, military entomologists. The US military has been involved in most of the major advances in preventing, treating, and understanding diarrheal disease. Cairo's NAMRU-3, the parent unit of Mark Riddle's humble container lab in Djibouti, has a four-star antidiarrheal pedigree. Its first director, Navy Captain Robert A. Phillips, figured out that adding glucose to rehydration fluids enhances intestinal absorption of salts and water. This meant rehydration could be achieved by drinking the fluids rather than making one's way to a clinic to have them administered intravenously. This has been a lifesaver not only for people who fight in remote, medically underserved areas but for people who live there. A 1978 *Lancet* editorial called Phillips's discovery "potentially the most important medical advance this century."

The full name of Riddle's study is Trial Evaluating Ambulatory Treatment of Travelers' Diarrhea (TrEAT TD).* "Travelers' diarrhea" is another catch-all term. Most of it—at least 80 percent—is bacterial, with 5 to 10 percent viral (vomit typically joining the waterworks here) and a miscellaneous percentage from protozoa like amoeba or giardia. All of it is caused by contaminated food or water. There used to be a separate category called "military diarrhea" (*military* referring to the patients, not the explosive nature of their evacuations), but if you look at the responsible pathogens, the breakdown is almost the same. Military diarrhea is travelers' diarrhea, because service members are travelers—in places where you don't want to be drinking the water. A survey conducted by Riddle, David Tribble,† and others with the US Naval Medical

---

* By the fair play rules of acronyms, this should be TEAT TD. Never mind, though. It's hard enough for diarrhea researchers to get the respect they deserve without bringing teats on board.

† Here is my diarrhea research statistic: When you are communicating with a pair of diarrhea researchers named Riddle and Tribble, there is a 94 percent chance you are going to slip up and refer to one or both of them as Dribble.

Research Center revealed that from 2003 to 2004, 30 to 35 percent of military personnel in combat in Iraq experienced situations where they lacked access to safe food and water. In the early days of a conflict especially, combatants are like backwater backpackers, crapping in the dirt and waving the flies off whatever food the locals are peddling. In that same survey, 77 percent of combatants in Iraq and 54 percent in Afghanistan came down with diarrhea. Forty percent of the cases were serious enough that the person sought medical help.

For every person who shows up at morning sick call, four tough it out. Riddle would like to know why. The average bout of travelers' diarrhea lasts three to five days. Why endure this, when some of the new antibiotics, Riddle's data show, can have you back to normal in four to twelve hours? He's been asking around, mostly at mealtimes. The tables in the hangar-size Dorie* are arranged church basement–style, in long rows, so there's always a friendly stranger across from you or at your elbow, someone new with whom to chat about loose bowel movements while you eat.

Riddle gets right into it this morning with the man to his left. The uniform identifies the man as a Marine sergeant, last name Robinson. "I'm in the Navy," Riddle is saying, "and we're looking at

---

* Full name: the Dorie Miller Galley. It is unusual for the military to use a nickname when naming a facility after one of its own. When the man's full name is Doris, an exception is eagerly made. Doris "Dorie" Miller was a cook who showed commendable bravery during the Japanese attack on Pearl Harbor, so commendable that his name appears on twenty-three government and civic facilities, eight opting for "Dorie" and fifteen—including the US Postal Service—embracing the full Doris. The US Navy named a frigate after Doris Miller. Since most frigates omit first names, the Doris issue was easily skirted, or pantsed.

simplified treatment regimens for travelers' diarrhea. We're finding that a single dose of antibiotic and an anti-motility . . . "

Robinson looks up from his eggs. "Anti—?"

"Like Imodium," I offer. "Stops you up."

"Oh, absolutely not. You do *not* want to mess with Nature like that." Robinson has the booming vocals and commanding bullnecked air of the actor Ving Rhames. One imagines Riddle going straight over to the lab after breakfast and tossing his data in the trash—*What was I thinking?*

"You have something bad in you, bad water or what have you? You got to pass it." It's like discussing diarrhea with the Giant Voice. "Defeat the purpose if you mess with that."

We've been hearing this a lot. People think diarrhea is the body's attempt to rid itself of invaders, or to flush out the toxins they produce. They won't take an antimotility drug like Imodium because they think it interferes with the purge. But diarrhea is not something you are doing to pathogens; it is something they are doing to you. In varied and dastardly ways. Shigella and campylobacter, two common causes of bacterial dysentery, wield a toxin-delivering "secretion apparatus"—a hypodermic-cum-bayonet that injects toxins into cells in the intestinal lining, killing them and causing the fluid inside them to spill out. That spillage is part of the watery-stool scenario, but there's more! With enough of those cells out of commission, the large bowel can no longer perform its duty as an absorber of water. Instead of food waste getting drier and more solid as it moves along the gastrointestinal tract, it stays liquid all the way along. The bacterium called enteroaggregative *E. coli* produces the same effect, in a different manner. It becomes a living cling wrap, a bacterial phalanx that coats the intestine and blocks absorption. Vibrio cholera and enterotoxigenic *E. coli*, or ETEC, inflict chemical weapons: Both produce a toxin that hijacks the pump that maintains cellular homeostasis. The commandeered pump begins pulling water out of cells faster than patients can replace it by drinking.*

Why do these creatures do this to us? Is there an evolutionary motive? Sure, says Riddle. There always is. By causing humans to produce liquid feces, feces that splatter and flow and coat a larger surface area, a pathogen speeds its spread. Cover the world! The bacterium that causes cholera is especially proficient. Cholera patients decant as much as five gallons of liquid a day. The efflux is so torrential that one of Dr. Phillips's Navy colleagues was inspired to invent the cholera cot, an army-style cot with a hole cut out under the buttocks. (Bucket sold separately.) The cots, still made today, allow patients to "go to the bathroom without leaving the bed," writes specialneeds cots.com, taking euphemism into the realm of quantum physics.

Besides, enteric bacteria are not easily flushed out. They've evolved ways to hang on in the deluge. ETEC—the bacteria behind as much as half of all travelers' diarrhea—are equipped with a hair-like grappling hook called a longus, which they use to pull themselves close to a cell wall. On receipt of a chemoelectrical signal from the cell, the bacteria sprout springy hairs called fimbriae, with suction cups at their ends. Your immune system, for its part, has more sophisticated defenses than simply hosing down the premises.

---

* The dose makes the poison. In small amounts, a mimic of the cholera/ETEC toxin is an effective treatment for constipation (in particular, the constipation that afflicts a third of irritable bowel syndrome sufferers). In 2012, Ironwood Pharmaceuticals released a synthetic version that was promptly forecast by one pharmaceutical market researcher to achieve "blockbuster status," and what could be more fitting for a constipation drug?

It starts cranking out specially designed antibodies. One might target the suction cups and keep them from adhering. Another might gum up the longus or disable the toxin.

Sergeant Robinson has nothing more to say about diarrhea, but he would like Riddle to have a word with the people responsible for the packet of toilet paper in the combat field rations, or MREs (Meals, Ready-to-Eat). "It's like this much."* He tears off a piece of napkin the size of a drink ticket. "To wipe your ass!" Riddle volunteers that Navy guys pack baby wipes. He may regret saying this, because Robinson counters that Marines just cut off a piece of their t-shirt. Which possibly sums up the whole Marine Corps–Navy relationship.

Riddle thanks Sergeant Robinson and makes to leave. He likes to get back to his quarters before 8:00 a.m., when the national anthems—first Djiboutian, then American—begin playing over the Camp Lemonnier public address speakers. All those outdoors have to stop what they're doing and stand respectfully until the music stops.† The Djiboutian national anthem is a melodic, sweeping num-

---

* I tried, but I cannot tell you who decided how much toilet paper to include in MREs, or how. But I can tell you a lot of other things about the TP, because I found the federal specifications, ASTM D-3905. I can tell you the required tensile strength, wet and dry. I can tell you the colors it's allowed to come in (white, dull beige, yellow, green), the minimum grammage and basis weight, the percentage of postconsumer fiber, the required speed of water absorption. And maybe that's our answer right there. Because if your anus is as securely clamped as the anus of whoever is in charge of "toilet tissue used as a component of operations rations," ASTM D-3905, you probably don't need much.

† Other bases require this at 4:30 or 5:00 p.m., when the flag is taken in. When the music begins playing, you stop what you're doing and face the flag. I was at Natick Labs when this happened. Without explanation, my hosts stopped talking, turned, and solemnly faced a display model of a new containerized latrine standing in the sight lines of the flag. Having heard about the horrors of open-bay toileting, it seemed wholly appropriate for us to direct some respect, however unintended, to the Expeditionary Tricon Latrine System.

ber, like the theme song to an old TV western. The whole thing isn't played, but it can seem that way should you be having some "postprandial urgency." Meals—particularly the big ones occasioned by all-you-can-eat chow-hall buffets—trigger the gastrocolic reflex, a major move-along of the contents of the large intestine. Ushering out dinner to make room for breakfast. If, on top of that, you have a touch of irritable bowel syndrome (IBS), there may be times when all the patriotism in the land won't keep you standing through the final bars.

During his years at NAMRU-3 headquarters, in Cairo, Riddle regularly got hit with diarrheal infections, a result of "sampling the fecal veneer" at local eateries. Irritable bowel syndrome is a well-documented, little-publicized aftermath of diarrheal infections—especially severe or repeated bouts. If you talk to people who've recently been diagnosed with IBS, about a third of them will say that their symptoms began after a bad attack of food poisoning. Defense Department databases reveal a five-fold higher risk of IBS among men and women who suffered an acute diarrheal infection while deployed in the Middle East. Even the Veterans Administration recognizes IBS—as well as a form of arthritis called "reactive"—as one of the "post-infectious sequelae" of enteric infections. If patients can show that the condition developed following an infection with shigella, campylobacter, or salmonella during deployment, they're entitled to benefits. Riddle estimates that the Defense Department could wind up spending as much money on these long-term consequences of food poisoning as it spends on post-traumatic stress disorder.

Why not prescribe antibiotics more widely? First, there's the issue of antibiotic-resistant strains developing, though this is less of a concern with some of the newer regimens that wipe out infections in a single day—likely not enough time for a resistant strain to evolve and thrive. More worrisome, perhaps, is recent research showing that the colons of overseas travelers who treat their diarrhea with antibiotics, particularly in Southeast Asia, tend to become colonized with two species of "bad" bacteria that they then carry home and can spread around town. Both bugs may inhabit a traveler's gut only briefly and cause no problems while they're there, but they are dangerous to patients with weak immune systems. Here again, with the newer single-dose regimens, it may not be an issue.

These are largely first-world concerns. The week I returned from Djibouti, the World Health Organization released a statistic for annual deaths from diarrhea worldwide: 2.2 million. The estimate for ETEC alone is 380,000 to 500,000 deaths per year. Children especially are at risk because they dehydrate dangerously fast. The Centers for Disease Control and Prevention puts the daily toll for deaths from diarrhea in children under five at 2,195—more than from malaria, AIDS, and measles combined. (The Gates Foundation is funding the Navy's efforts to develop an ETEC vaccine.)

Riddle traveled a lot in his twenties and recalls being hit by a realization. So much of people's lives—their opportunities, their health and longevity—comes down to where they were born. "It's so random," he says. We're over at his office, which is downstairs from his lab, in the same container. "It shouldn't be that way. It shouldn't matter where your parents happened to live." He pauses for a jet ripping through a takeoff. At certain times of day, you get this every few minutes. It's like having a desk under the tarmac at Heathrow. The commotion fades and Riddle resumes. "I went into medicine wanting to help the greatest number of people." And then, just when I thought he'd gone all earnest on me: "I happened to fall into diarrhea."

. . .

AT CAMP Lemonnier there is often a quicker route to where you're going, but to follow it you'd need to be shot. You'd need to scale a twelve-foot-high Cyclone fence and the barbed-wire Slinky along top of it, ignore the sign—Stop! Deadly Force Authorized!—and cross into the secure zone. Camp Lemonnier is the hub of counter-extremism activities in north Africa and Yemen. A fleet of drones resides in the zone, along with Navy SEALs and other Special Operations ghosts who pass in and out on their way from one classified gig to the next.

These are the people I want to speak with. I'm interested in diarrhea as a threat to national security. How would the takedown of Osama bin Laden have played out had one of the SEALs been fighting the forces of extreme urgency? How often is food poisoning the cause of a "mission fail"?

Yesterday I convinced the droll and adorable Camp Lemonnier public affairs officer, Lieutenant Seamus Nelson, to put a request in the daily email feed that goes out to everyone on base. (". . . Mary is looking for individuals who would be willing to share a story about how a case of diarrhea has impacted them while engaged in operations. . . .") Because really, how do you step into that conversation? The men of Special Operations are easy to spot—the beards, the build, the air of stony omnipotence—but they are not easy to talk to. They keep solidly to themselves. You don't find them in the bar or the Combat Café. No one from Special Ops showed up at the LGBT barbeque or competed in the Fourth of July Cardboard Boat Regatta. Nor will you find them at the gym. They have their own equipment and trainers inside the zone.

"They only come out to eat," says Riddle. We are sitting in Seamus's office, waiting to talk to one of the four people who replied to the now infamous diarrhea email.

Seamus nods. "And to steal our women."

The interviews have been scheduled back-to-back, one man coming in as another leaves, the Public Affairs container having taken on the quiet, hangdog air of a Catholic confessional. We just listened to the commanding officer of an inshore boat unit that protects Navy ships from USS *Cole*–type terrorist attacks in the port of Djibouti City. He demonstrated the maneuvers using Seamus's stapler as the "high-value asset" kept safe by a tape dispenser and a bottle of allergy pills, zigzagging across each other's paths. An inopportune bathroom break would leave the stapler vulnerable to attack. Even if crew stick to their posts, their vigilance is compromised; "illness preoccupation" is an overlooked military liability of diarrhea.

We heard a similar tale from a bombardier. On a long sortie out of Diego Garcia island, the only crew member capable of operating the plane's defensive equipment abruptly left his post to use the chemical toilet—while flying over Taliban-controlled Afghanistan. On the return flight, a faulty seal combined with the pressure differential between the toilet's tiered chambers caused the contents to spew into the crew cabin. "Be assured," he deadpanned, "this blue-brown precipitation affected the navigator's ability to concentrate on his duties."

Our 3:30 is retired from Special Operations, now working as a contractor. A tattoo on the inside of one forearm depicts a pair of crossed metal objects that I can't identify but guess to be martial arts weapons of some variety. When I ask what his job is, he answers cryptically, "I fix things." I take this to be a euphemism for some unspeakable niche calling—eliminating witnesses, disposing of bodies, God knows. Subsequent conversation reveals that the man is, in fact, a mechanic. *He fixes things.* The objects in the tattoo turn out to be pistons.

The mechanic was hit with diarrhea every time his team

deployed. Because of this, he was never assigned any "long-range sur-veillance," meaning counter-terrorism missions deep into insurgents' turf. These missions, he says, entail hiding out in a hole,* watching a particular spot—say, an intersection: who comes and goes, how many trucks drive through, at what time of day.

I nod, but don't entirely understand. "To find out . . . ?"

"Do we aim a bomb here."

"Ah." Silly me.

I ask the Special Operations mechanic whether he knows of a vital operation that might have been compromised because someone got a vicious case of food poisoning. He dismisses the very idea. "The guys they select for this type of work? They don't have these types of problems. They're selected for a reason."

After he leaves, Seamus turns to us. "Wow, do you think that's part of the screening for Special Operations? Give you some bad food, see how you do?" He's joking, but in fact 20 percent of the population are what Riddle calls "nongetters": people who can eat ceviche from street vendors, drink the water, never get sick. It would certainly be an asset. Riddle wonders whether Special Operators take antibiotics or Imodium prophylactically, just in case, before critical missions. Or are they just suffering in silence? The Camp Lemonnier Special Ops doctor—they have their own, natch—talked about the men's reluc-tance to seek medical help lest they lose their Special Operator status.

Riddle and I have a lot of questions. Alas, no one from Special Operations replied to the diarrhea email.

---

* Tips for hole-living: Double-bag your peanut butter sandwiches in gallon Zip-locs, as the bags serve double-duty as the toilet. Bring cat litter to put in the bags in case diarrhea strikes, which it does reliably enough that the man who told me this, an air strike controller just back from Niger, was confronted by his commander wanting to know why Special Operations Command was requisitioning kitty litter.

Perhaps a second email is in order, this one offering compensation. Riddle advises against it. He says people will make up a story to get the cash. He has had men sign up for the diarrhea study, go into the bathroom, and hand him a Commode Specimen Collection tub with a perfectly formed turd inside.

"Also?" Seamus again. "I'm done sending out PSAs about diarrhea. I'm set." He got some blowback from Combined Joint Task Force–Horn of Africa headquarters regarding appropriate content for base-wide email.

Chow is my one shot.

SEAMUS NELSON is six foot three. When he extends his neck to its full reach, his head is like a periscope. It's up now, surveying a sea of clean-shaven, supper-chewing heads in the Camp Lemonnier dining facility. He's scanning for facial hair. Only two categories of men here are allowed to wear beards:* Special Ops and civilian contractors who want to look like Special Ops.

---

* The clean-shaving rule began with the gas attacks of World War I. Whiskers compromise the airtight seal of a gas mask. (Special Operators are exempt because they may need to blend in with bearded Muslim locals, and because they're Special.) There were also some hygiene concerns. In 1967, the Department of the Army undertook an investigation entitled "Microbiological Laboratory Hazard of Bearded Men." To see whether bearded sixties bio-warfare lab workers might be putting their family members at risk via "intimate contact," the researchers fashioned some human hair beards, contaminated them with deadly pathogens, and attached them to manikin heads. The heads then became intimate with some chicks. "Each of three 6-week-old chickens was held with its head alternately nestled in the beard and stroked across one-third of the beard (one chicken on each side and one on the chin)." When the beards were washed according to lab safety protocol, none of the nine chicks exposed to the highest concentration of the virus became infected. The heads with unwashed beards, however, transmitted deadly disease to the chicks with whom they'd been intimate. The chicks died, and the heads were never really the same after that.

"There's your guy." The neck now retracted. "Far corner by the door."

Riddle and I rise from our seats. We saw this man yesterday, coming out of the tactical shop. Even without the beard, you'd know he's one of them. There are men who attempt to broadcast toughness by what they wear or drive or have tattooed on themselves. And there are others, like this man, who do nothing to cultivate or consciously project it, and yet it is obvious. It accretes naturally of the things they've experienced.

Besides, I saw him go into the secure zone.

This promises to be awkward. It's not just the topic. It's how people like this make you feel: the sudden reveal of your smallness and the inconsequential preoccupations of your existence. What could be smaller than writing about diarrhea? And how to explain why I've singled him out?

"Seamus, come with me. Introduce me."

Seamus peels an orange, one long strip spiraling down to his tray. "I don't know, Mary. We weren't trained on this in public affairs school."

I collect my notepad and tape recorder.

"Hang on." Seamus, clearly stalling, wipes orange from his fingers, one at a time. "I'm going to be shaking his hand. He'll kill me." He lowers his voice: *You got me sticky."

I stand up. Seamus makes a brief, warbling unhappiness sound and pushes back his chair.

We cross the cafeteria, nervous middle-schoolers at the dance. The man sees us but does not alter his expression. We stop a couple feet back from the table. Some kind of attitudinal concertina wire. Seamus plunges ahead. "Mind if we join you for a second?"

The man takes hold of the sides of his meal tray. "I'm done."

"We . . ."

"I'm leaving."

Seamus keeps paddling. "Do you have time for a quick question, what line of work are you in?" *What line of work are you in!* I adore Seamus Nelson.

The man glances at Seamus, at me, and back to Seamus. "Who are you." Said like something thrown.

"I'm in Public Affairs, and this is an author. She's working on a chapter for a book, and she's specifically focused on how diarrhea impacts a mission . . ."

This is my cue. I'm going to assume the man is Special Operations, and that he knows we know. "I was wondering whether you might ever have been in a situation where . . . in a critical mission that . . ." I back up. "Well, because diarrhea is looked on as sort of a silly—"

"It's not."

He speaks softly, and what he says next I can't quite make out. Something about being curled up in a hole in the fetal position. He says that where he just got back from, some unnamed "out station" in Somalia, it hits everyone. This is probably not exaggeration. In Riddle's survey of diarrhea in Iraq and Afghanistan, 32 percent of respondents reported having been in a situation where they couldn't get to a toilet in time. And Special Operators in the field get sick twice as frequently as everyone else.

His name, he says, is Carey. He invites us to sit down. I place my tape recorder in plain view—that is to say, in plain view of anyone on my side of the table. That is also to say, behind the condiment caddy.

I need Carey to set the scene. "What if you . . . I mean, what if *someone* were a sniper, and they're in a hide for . . . well, how many hours would it be?"

"Depends on the mission. You're watching for something to happen that might not happen."

"Right, and most likely you're out in some village, and you've had to be eating stuff that's not prepared as hygienically as—"

"Goat," he says. I had heard a story earlier about a goat meal in rural Afghanistan. It contained the phrases "singed hair" and "otherwise uncooked." Unsanitary conditions, Carey confirms, are a given. "Unfortunately, we don't fight in first-world countries."

Carey says he does not, as Mark Riddle had heard some men did, take antibiotics or Imodium prophylactically before the mission or after the goat. He takes one precaution. It is a strict rule among Special Operators. "You go to the bathroom before going into a danger situation." There has been no shift from the gravely quiet tone with which Carey has been speaking. Nonetheless, Seamus blurts, "Kind of like a road trip with the family, and Dad's like, 'I don't care that you don't need to go.'"

On a family road trip, no one has you in the sights of a semiautomatic rifle while you squat in the dirt. Historian of military medicine A. J. Bollet quotes a letter written by a Civil War soldier who explained that an unwritten code of honor forbade the shooting of a man "attending to the imperative calls of nature."* In the war on terror, there's no such etiquette.

I'm still trying to get Carey to tell the story of a specific high-stakes operation. "Have you have been in a situation where you've been—"

"Inabilitated?" I like this: a combination of *inability* and *disabled*.

---

* Dale Smith, a historian of military medicine at the Uniformed Services University of the Health Sciences, is dubious. Bollet, he says, drew the conclusion from one man's story. Certainly no such etiquette prevails among military historians, who take any opportunity to shoot each other down.

"Yes. I have been inabilitated because of food sickness." Carey leans back, one arm along the back of the adjacent chair. "I'm not sure what you guys want from me."

Seamus tries to help. "Can you walk us through the story. You know, like: *There I was . . .*"

Carey isn't going to supply the There-I-Was. "I have many stories where I've soiled my pants on missions. In Iraq, I've soiled my pants. In Afghanistan, I've soiled my pants." No one stays back or leaves to find a toilet once an operation is under way. Diarrhea cannot be a "kill stopper."

"And then what happens?" Seamus leaning forward like a kid at story hour. "You go on to . . . do the job?"

"There's no other option. I mean, it's kind of a life or death thing. So." He shrugs one shoulder. "You go. Worry about it later. As long as you walk out and the mission is accomplished. And that's about as specific as I can get."

I tell him about Mark Riddle's TrEAT TD study. "You should bring along a single sixteen-hundred-milligram dose of rifaximin and a bottle of Imodium."

Carey holds my gaze for a moment. "What is the objective here?"

I restate my mission. I show him my notebook, open to the page where Mark Riddle is describing what, for the purposes of his study, constitutes diarrhea ("has to be pourable or take the shape of the container").

"Well, you're in the wrong place, Mary." Carey tells me to go down to Somalia. Yes, let's picture it—middle-aged American with her cork-bed comfort sandals and wheelie bag, wandering the desert redoubts of the local al-Qaeda affiliate. *Yoo-hoo! I'm looking for the Navy SEAL safe house?*

"You could get yourself down there if you wanted to. It's not dangerous." He pushes two fingertips through the curl of his beard. "Well, it's a little dangerous."

Carey apologizes for the frosty reception earlier. "I thought you guys were NCIS." Naval Criminal Investigative Service. "You scared me."

CAREY IS right. People don't get diarrhea by eating at Camp Lemonnier. They get it by "eating on the economy": the Special Operators get it in remote villages, and everyone else gets it by going in to Djibouti City for a change of pace from spaghetti and Taco Tuesdays. Like you on your Mexican* holiday, they ingest contaminated tap water or food that's been sitting out unrefrigerated. Before a downtown suicide bombing caused the base to be put on restricted liberty, a month before I arrived, Riddle was seeing two dozen food poisoning cases a week. During the past month, since everyone's been confined to base, only one person—the guy who found a restaurant

---

* How did Mexico become the poster child for travelers' diarrhea? One hypothesis, mine, points a finger at the godfather of diarrhea research, Herbert DuPont. For almost thirty years, DuPont ran studies out of Guadalajara, Mexico. If you plug "Guadalajara" and "diarrhea" into the PubMed database, you get forty-five journal articles and a persuasive argument for changing your holiday destination to Switzerland. ("Enteric pathogens in Mexican sauces in popular restaurants in Guadalajara . . ."; "Coliform contamination of vegetables obtained from popular restaurants in Guadalajara . . ."; "Coliform and *E. coli* contamination of desserts served in public restaurants in Guadalajara . . .")

There has been at least one well-intentioned effort to clear Mexico's name. The author of a paper in *California Medicine* had read that Mexicans often get travelers' diarrhea when they visit California. She wondered if perhaps the stress of travel, rather than poor sanitation, was to blame. She interviewed 215 foreign UCLA freshmen and 238 American freshmen about "changes in frequency and consistency of stools." None of the foreign students appeared to have had travelers' diarrhea, though it was difficult to tell because many "did not understand the interviewer's terms." You can see where "watery stool" or "explosive diarrhea" might be confusing, frightening even, for the non-native speaker.

that delivers—has stepped through the door. Riddle catches up on paperwork. The lonely diarrhea researcher.

The Camp Lemonnier galley goes to lengths to keep bacteria away from the food. The entry hall is flanked by rows of knee-operated hand-washing stations, with pole-mounted Purell dispensers at the end of these. All well and good, but here's what really matters. First, the cooks and prep crew wash their hands after they go number two. So if any of them has diarrhea, that person isn't spreading the bacteria to food that then sits out at room temperature, allowing said bacteria to multiply to levels at which they cause illness. And second: There are no flies in the Dorie Miller Galley. Since the dawn of air-conditioning, military chow halls have been sealed. No one needs to open a window, and no one can.

It was as a result of this connection—fewer flies in the mess equals fewer gastrointestinal infections—that the filth fly was originally busted as a vector of disease. In 1898, during the Spanish-American War, a trio of army physicians, including the illustrious and eponymous Walter Reed, were called to Cuba to investigate an outbreak of typhoid fever that was killing one in five US troops. (It was Reed's medical sleuthing that proved it was mosquitoes, not bad air or unclean bedding, that transmitted yellow fever.) Straightaway, the team noticed that the infection rate was lower among the officers whose mess tents had screens to keep bugs out. It also varied by the different camps' methods of "disposing of the excretions." Open pit latrines were associated with higher rates, possibly because, as Reed's team wrote, "flies swarm over the infected fecal matter."

Reed had his two suspects—flies and the bacteria-laden feces they feed on—but no smoking gun. Flies don't bite. How were they transmitting the pathogen? One fine day Reed's gaze fell upon a fly walking around on the soldiers' food. Looking more closely, he noticed white powder on its little hairy legs. Where had he just seen a white powder? The pit latrines! Soldiers had been sprinkling lime in an attempt at camp sanitation. The flies' feet were delivering bacteria from shit to stew. They're what's called a mechanical vector. Ten *Salmonella enterica* serovar Typhimurium isolates multiplying in a pot of beans in the warmth of a Cuban noon will be a million by dinnertime.

Reed's legacy lives on in the Entomology Branch of the Walter Reed Army Institute of Research, my next destination. Military entomology runs the gamut you'd likely expect: killing disease-carrying insects, keeping them away from soldiers, and creating vaccines and treatments for the times when neither of those can be managed. In the case of filth flies, something less usual has been going on. Unlike in most military entomology war stories, the insect this time is the hero.

9 The Maggot Paradox

Flies on the battlefield, for better and worse

I N A MEMORABLE CARTOON from my formative years, a well-dressed man with a goatee is seated at a restaurant table across from a fly. It's a giant fly, a fly large enough to fill a dining room chair much the way a person would. The man addresses a waiter. I'm paraphrasing here. "I'll have the gazpacho, and some shit for my fly." It was a commentary on flies, or perhaps an observation on the odd human habit of elevating nutrient intake to social ritual. Or maybe just this: No matter how fond you are of a fly, dining out together is going to be awkward.

And the cartoonist only drew the half of it. Because flies have no teeth, they must first liquefy what they plan to eat. (Or order the gazpacho.) This they do by applying their digestive enzymes outside their body. The process was captured on film and included in the 1940s British Army hygiene filmstrip *The Housefly*. "Their vomit is puddled about your food to make a kind of porridge," says an incongruously posh-sounding narrator, "which the fly then sucks up." Technical Guide No. 30 (*Filth Flies*) of the US Armed Forces

Pest Management Board would also have you know that "flies further contaminate food by defecating on it while they feed."

No flies of any size are eating at Mi Rancho Mexican restaurant in downtown Silver Spring this evening, but some fly biologists are here, and that can be equally disquieting. We're talking now about those out-of-body digestive enzymes. A researcher I had spoken to the previous week referred to the salivary glands, not the stomach, as the source. For clarification, I have turned to one of my dinner companions, George Peck, resident filth fly expert at the Entomology Branch of the Walter Reed Army Institute of Research (WRAIR), just down the road.

"I think it's both," Peck is saying. "They vomit up enzymes from the crop along with the saliva and let it—"

"Are you all done here?"

Peck looks up to acknowledge our waitress. "I am, thanks . . . and let it fall onto the food."

With George Peck, the topic of flies and their unusual physiology doesn't elicit disgust. Awe, mostly. I have heard him marvel at the sensitivity of the fly's body hairs, how they enable it to detect the bow wave of an approaching hand and lift off in the split second before contact is made. He talks about the halteres, tiny gyroscopes that enable the fly to hover or change direction "faster than the fastest flight computer on any jet."

Less awesome: Researchers in Japan established that the strain of *E. coli* known as 0157:H7—deadly outbreaks of which periodically make headlines in the United States—thrives in housefly mouthparts and frass.* Bacteria on or in filth flies have been shown to transmit typhoid fever, cholera, dysentery, and a whole wet bar

---

* Insect shit.

of lesser diarrheal infections. (Both houseflies and blowflies fall under the grouping "filth flies.") British researchers documented a close association between filth fly populations and cases of food poisoning from the campylobacter bacterium, both of which peak during the warmest months. (The English used to speak of "summer diarrhea"—loose stools and cramping having joined warm nights and fireflies as hallmarks of the season.) In a 1991 study, an Israeli military field unit that undertook an intensive filth fly control program saw 85 percent fewer cases of food poisoning than a similar one that did not.

The Armed Forces Pest Management Board's filth fly technical guide includes a figure for the number of times in twenty-four hours that a single fly vomits and defecates on its food after a controlled feeding of milk. The figure, a range from 16 to 31, was arrived at not by staying up all night watching but by counting "fecal spots" and "vomit spots" (the latter distinguishable from the former by their lighter color). The reader is invited to speculate about the number of "spots" on food in a military chow line in the era before sealed dining facilities. Fly infestation in the mess halls of the Vietnam War, the guide relates, was so intense that "it was difficult to eat without ingesting one or two . . ."

Infestations still happen, mostly in the rough and not entirely ready first few days or weeks of a war. Early on, weapons and ammo take priority over latrines and refrigeration units in terms of what supplies get shipped. During the first Gulf war, Marines arrived in the region via the port of Jubail, where the Saudis housed them in a warehouse. "We had ten thousand Marines and two squat toilets," recalls Joe Conlon, a retired Navy entomologist. The toilets soon clogged and sewage ran in the streets. Meanwhile, with no refrigerated storage, pallets of produce began piling up on the dock in the

100-degree heat. Thousands of flies converged. Conlon estimates 60 percent of the Marines got sick.

Historically, battlefields were even worse. Combat is a filth fly cornucopia—a bounty of rotting organic matter to eat, to lay eggs in, to nourish the offspring. On Pacific islands during World War II, says the Armed Forces Pest Management Board guide, "flies developed in corpses on battlefields and excrement in latrines to levels beyond modern comprehension." A similar scenario developed in the aftermath of battle in El Alamein, Egypt, prompting officers of the British Eighth Army to mandate fly death quotas—each soldier responsible for killing at least fifty flies a day. During the Vietnam War, corpses became so heavily infested with maggots that pesticides had to be used inside body bags.* In Conlon's camp on the Kuwait border, accumulating garbage exacerbated the problem. The Marines weren't allowed to burn it—the normal disposal strategy—because the fires would give away the camp's position. (The garbage eventually became part of military strategy. It was hauled away under cover of darkness and burned at a distant site, to trick the Iraqis.)

Nowhere was the filth fly situation more dire—or perhaps just more memorably documented—than in the American Civil War. "Few recruits bothered to use the slit trench latrines . . . ," wrote Stewart Marshall Brooks, in *Civil War Medicine*. "Garbage was everywhere . . . [alongside] the emanations of slaughtered cattle

---

* Post-Vietnam-era mortuary practice forbids this, as the pesticides could interfere with the chemical and genetic analyses done as part of an autopsy. Also verboten in morgues: electric fly zappers. They cause the flies to explode, scattering their DNA and the DNA of whatever bodies they've been crawling on. Military morgues rely on "air curtains" to keep flies out. The air curtain is a high-tech version of the "fly curtain," the beaded strands that hang in doorways in Middle Eastern homes, allowing breezes, but not flies, to pass. Who among the thousands of youthful 1970s doofs who hung these in their bedrooms had any clue as to the beads' provenance as fly control? Not this doof.

and kitchen offal." Entomologists Gary Miller and Peter Adler, in a paper on insects and the Civil War, quote a letter by an Indiana infantryman describing the scene: "The deluge of rain which had fallen . . . soaked the ground until the whole face of the earth was a reeking sea of carrion. . . . Countless thousands of green flies . . . were constantly depositing their eggs . . . which the broiling sun soon hatched into millions of maggots, which wiggled until the leaves and grass on the ground moved and wiggled too."

You can imagine what might happen to the open wounds of a soldier lying on a battlefield for any length of time. Most likely you would be wrong.

THE SOLDIERS, two of them, are not named, nor is the battlefield on which they were hit. We know that it happened in France during World War I, sometime in 1917. We know that it wasn't winter, because the men arrived at an army hospital having lain "in the brush" for seven days. And because it was fly season.

> On removing the clothing from the wounded part, much was my surprise to see the wound filled with thousands and thousands of maggots. . . . The sight was very disgusting and measures were taken hurriedly to wash out these abominable looking creatures. Then the wounds were irrigated with normal salt solution and the most remarkable picture was presented. . . . these wounds were filled with the most beautiful pink granulation tissue that one could imagine.

That's US Expeditionary Forces surgeon William Baer relating the story of how he came upon the unseemly idea of intentionally infesting wounds with maggots to help them heal. Filth fly larvae—blowfly maggots, most notably—prefer their meat dead or decaying. When the meat is part of an open wound, the act of eating performs upon the meal a kind of natural debridement. Debridement—the removal of dead or dying tissue—fights infection and facilitates healing. Because dead tissue has no blood supply and thus no immune defenses, it's easily colonized by bacteria. This encourages infection of the healthy tissue and inflammation, which slows healing.

Baer was impressed that the soldiers had no fever or signs of gangrene. The mortality rate from the type of injuries the men had—compound fractures and large, open wounds—was about 75 percent with "the best of medical and surgical care that the Army and Navy could provide." In 1928, a decade after the war had ended, Baer summoned his courage and experimented on civilians. His inaugural patients were children, four of them, all with recurrent bone infections from blood-borne tuberculosis, a condition that antiseptics and surgery sometimes failed to quell. Raymond Lenhard, the author of a biographical monograph on Baer, recalled hearing the great surgeon tell the story. Lenhard had been a student of Baer's at Children's Hospital School in Baltimore and, reluctantly, a dining companion. ("Often during lunch he would make us lose our appetites.") Using the offspring of blowflies trapped near the hospital, Baer "loaded up" a wound and proceeded to watch the results. After six weeks, the wound had healed. As did the wounds of the other three children.

What sort of person experimentally infests a child with maggots? A confident sort, certainly. A maverick. Someone comfortable with the unpretty facts of biology. Someone who is perhaps himself an unpretty fact of biology. "The Chief was overweight, breathed audibly, and snorted in the fashion of a tic," wrote Lenhard. Baer would sometimes go from operating room to lecture hall without changing, delivering his talks in baggy, bloodstained surgical trousers. He bred Chow Chows at his home, bringing yet more snorting and audible breathing to the Baer household.

Beneath the earthy exterior, Baer was an exacting and dedicated practitioner. He considered his "maggot treatment" far less abhorrent than the alternative: amputation. To Baer, the removal of a limb was "the ultimate in destruction," wrote Lenhard, showing a flair for video game marketing eighty years premature.

So impressed was Baer by the work of his larval "friends" that he designed and built a thermostat-controlled wood and glass fly incubator at the hospital. Only thrice in an ensuing eighty-nine cases did the maggots fail and the patient succumb to infection. Fearing that the larvae may have introduced the offending bacteria, Baer devised a protocol for raising sterile specimens. Remnants of his technique live on today at Monarch Labs, in Irvine, California. Their Medical Maggots are also sterile, as required by the Food and Drug Administration (FDA), which in 2007 approved live blowfly larvae as a medical device.

While the majority of modern "maggot therapists" treat the hard-to-heal foot ulcers of diabetics, WRAIR's George Peck has been seeking to take medicinal maggots back to their roots in the military. In 2010, he was funded for a study looking into the efficacy of blowfly larvae in treating chronically infected IED wounds. More recently, Peck received a grant to genetically modify blowfly maggots such that they produce antibiotics. Though maggots already prevent infection, these "supermaggots" could be tailored for specific bacterial infections.

Peck offered to hatch a "clutch" of maggots for me, taking care to time things such that when I arrive at his and his wife's home for dinner, the larvae will be the size of Medical Maggots at the time they're released in a wound (about two millimeters long). I don't have any wounds. Just questions.

GEORGE PECK and his future wife, Vanessa, worked together in the basement insectary at WRAIR. An insectary is a facility for rearing insects—insects used, in this case, for testing vaccines and repellents against whatever has been lately plaguing troops. Vanessa cared for a colony of sand flies,* while George was down the hall with his filth flies. It's a setting that might dampen the ardor of another pair, but Peck remains besotted. You hear it behind his words when he talks about her. Peck is a man easily taken by emotion. At Mi Rancho a few nights earlier, as we were getting ready to leave, the topic turned briefly away from flies. As I rose from my chair I heard Peck say, to no one specific, "I just love bees." The word *love* breathy with feeling.

Peck abandoned a career in solar physics, because he felt it was taking him too far away from the natural world. He and Vanessa share their home with more of that world than most. They keep as pets a tarantula (Henrietta) and a small community of Madagascar hissing cockroaches. Like William Baer, Peck is a man some might

---

* Tobin Rowland, the man who now holds the job, gave me the WRAIR Insect Kitchen recipe for sandfly larvae food. Mix rabbit feces, alfalfa, and water, and pour into nine large round pans. Soak for two weeks, or until mold covers entire surface, yielding what WRAIR entomology director Dan Szumlas calls a "lemon-meringue feces" appearance. Let dry and grind. Rabbit dung is used because it smells better than cow dung, not because it's cheap. Rabbit turds are more expensive than rabbits. WRAIR's supplier, which holds a monopoly by virtue of no one else's having wanted or thought to compete, charges $35 a gallon.

find eccentric, but those who know him even slightly can see that it all comes down to a generous and open heart.

Vanessa clears the dinner plates while I finish my wine. The children are doing homework in the living room. George sets a glass dessert dish in front of me. *Chocolate pudding*, my brain offers optimistically, but it's not that. It's raw liver.

"These are about one day old." Peck points out a cluster of maggots, maybe twenty or thirty, feeding side by side, packed in close. They're easy to miss, because all that can be seen of them is their tail ends. Insects take in oxygen through openings in the exoskeleton called spiracles. In the larvae, these are, specifically, anal spiracles. On top of its other charms, the maggot breathes through its ass. It is a handy evolutionary adaptation if, as Peck puts it, "you spend your whole day with your head buried in slimy dead flesh." Compared to lungs and a diaphragm, it's an inefficient system, which is one reason the Insecta class never evolved to be as large as Mammalia. Having several minutes ago viewed a fly under George Peck's home microscope, I assure you that's a good thing.

William Baer likened clusters of feeding maggots to litters of puppies. "So voracious are they in their struggle for food that they will stand upright on their heads with their tails in the air, as puppies do . . . around a basin of food where the basin is too small for the number of puppies." Baer had dogs on the brain. To me, they look like a set of tiny accordion buttons being played by some ghost polka virtuoso. The important thing, especially for someone being treated with them, is that they don't look like maggots. So if a patient peeked beneath the trademarked Monarch Labs LeFlap dual-layered maggot cage dressing, he would not be slammed with a squirming Halloween horror visual.

Peck transfers three outliers to the tip of my index finger. They rear up and wave their heads like happy *Sesame Street* puppets. Peck says they're searching for food. Now two are lifting the third up above them. They remind me of jubilant teammates after a sporting victory.

Peck isn't reading joy in the scene. "They do cannibalize," he says gently.

Upon closer inspection, they are, yes, attacking—*eating!*—their clutch mate. They were away from the liver for maybe two minutes! The maggot lives to eat. That is what it does, all it does, for the four or so days before beginning the energy-intensive, deeply sci-fi project of rearranging itself into a fly.

Peck puts a maggot under the microscope he has set up on the kitchen table, so I can get a closer look at the mouthparts: the showpiece of maggot anatomy. They are rasping, curved scythe-like things. They're the only piece of the maggot formed from chitin, hard and brown in contrast with the creature's moist, pale, flexible self. Fortunately for maggot debridement therapy patients, the tissue deep inside a wound—dead or alive—has no sensory nerves; those are up in top layers of skin. Provided the Medical Maggots recommended "dosage"—5 to 8 maggots per square centimeter of wound surface—hasn't been exceeded, there should be enough dead tissue to go around, and no ravenous maggot will shift its gaze to live skin.

"Those little mandibles," Peck says as I look through the eyepiece, "can do what no surgeon or scalpel can do. No robotic laser can bend its light into a hidden crevice from an IED blast like that can. *That* is the master surgeon." If you want to destroy every last bacterium and shred of dead tissue, a maggot is your man. He's a small man, though, so it takes a while. A course of maggot debridement therapy—up to six rounds of fresh larvae—may take weeks. Whereas surgical debridement can be done in a matter of hours. And if a patient's immune system is healthy, as a young soldier's tends to be, it's not necessary to get every last cell of bacteria and necrotic flesh.

But Peck never suggested that maggots be used for the initial debridement of a blast wound. For military personnel, maggots would come into play further down the road, should a recalcitrant infection set in—some antibiotic-resistant strain, perhaps something exotic and stubborn that lurked in the dirt that was blasted so deeply and voluminously into the wound. These complications set in often enough that Peck received military funding for a rodent study to measure the effectiveness of maggot debridement therapy for soil-infected IED wounds. The experimental protocol presented challenges. It required Peck's team to surgically mimic, in a rat, the typical injuries caused by a bomb blast. To meet the requirements of the animal review board—and Peck's own personal ethics—no part of the process could be painful to the rat. The nerves supplying feeling to that portion of the body had to be identified and severed.

Peck's funding was not renewed, for reasons that are fairly easy to guess at. Modern hospital culture is technology-driven and forward-looking. To those unfamiliar with the studies and success rates, maggot therapy sounds primitive, anachronistic. Peck recalls presenting some promising preliminary findings to a roomful of colleagues and listening while a disapproving colonel talked about the thirty years of advancements he'd seen in his time at WRAIR. The man shook his head. "And we're using *maggots*."

A 2012 survey of US Army physicians suggests that the colonel's opinions are not those of the majority. While only 10 percent of those surveyed had prescribed maggot debridement therapy, 85 percent felt that having access to practitioners would be a good resource. Their reservations were mostly just practical: They didn't know

176 ■ Mary Roach

where to obtain the maggots or how to use them or what the billing code might be.* In a smaller survey, practitioners voiced concerns that the medical facility they worked for wouldn't allow maggots, and that patients would be likely to balk.

They are wrong about the patients. The surgeon who runs the Southern Arizona Limb Salvage Alliance,† David Armstrong, has applied maggots to more than a thousand patients. "I can count on one hand the number of people who have refused," he told me. The Medical Maggots FDA approval summary cited a "complaint and adverse event" rate of 1 percent, many of these occasioned by "late or lost" (or perhaps hurled into a Dumpster by the driver) FedEx shipments. The yuck factor of these wounds—and their resistance to more conventional treatment—well overrides the yuck factor of hosting live blowfly larvae. Also, Medical Maggots are less off-putting than you might imagine. Straight out of the vial, they're the size of cupcake sprinkles. When they're not eating each other alive, they're kind of adorable. They move like inch-worms, like something you might see humping along the pages of a children's book.

"People take an interest in the cute little guys," Armstrong said, quickly amending his statement with ". . . and gals." You mean, I asked him, the way one might follow the progress of seedlings one has planted, or guppies one is raising? "Exactly," he said. "And then, in turn, the progress of the healing going on. It's hard to describe it, but the larvae draw people into the wound emotionally." Medical Maggots patients, some anyway, are sufficiently positive and cavalier

---

* Medicare reimbursement code for maggots: CPT 99070.

† I am inclined to like a man who creates—for a medical practice that specializes in bowl-shaped, moist red wounds—the acronym SALSA.

about their infestations that they go around wearing Monarch Labs t-shirts that proclaim, "Maggots on Board!"

Hospital staff are less charmed. "A lot of doctors and nurses find it repulsive," Armstrong told me. Colonel Pete Weina, former director of the Complex Wound and Limb Salvage Center at WRAIR and now their chief of research programs, agrees. Around 2009, Weina had a William Baer moment. "I had a patient who'd passed out in an alley and flies had come by and laid eggs in his wound. The nurses were all, 'Oh my God, this is terrible, get the maggots out of there!'" Recalling what he'd read about the blowfly larvae's talent for debridement, Weina improvised a cage dressing to keep them from straying and left them in. The wounds healed nicely, but Weina backed away from the practice. "The entire hospital was pretty much grossed out by what I was doing."

While not discounting what he calls "the gross factor," George Peck sees cost as the main hurdle. How is it, you might ask, that maggots are more expensive than surgeons? It's not the creatures themselves; a vial of Monarch Labs maggots is priced at $150. It's the time demands on medical staff—staff who have to be trained to monitor the maggots and change the dressings. Peck shows me a second bowl of liver and maggots, hatched two days earlier. "See how foamy and goamy it is in there?" With, say, a hundred maggots, he explains, the breathable mesh of the cage dressing quickly becomes nonbreathable. The larvae suffocate. The nurses are repulsed.

Changing a maggot dressing is trickier—and creepier, and *goamier*—than changing other kinds of wound dressings, because you are also changing the insects. Each dose must be completely wiped out—literally, with a piece of gauze—before the next is introduced. Overlooked maggots that continue growing will soon be gripped by an urge to pupate. After a few days of gorging, fly larvae abandon the juicy chaos of their childhood home and set out to find a dry, quiet place in which to build a cocoonlike "puparium" and become a fly.

There is an understated line in the Medical Maggots package insert: "Escaping maggots have been known to upset the hospital staff . . ." One, they're maggots. Two, they're about to be flies. Flies in the medical center. Flies in the operating room. Landing on open wounds. Vomiting and defecating. Moving on to other wounds, spreading the antibiotic-resistant pathogens they've picked up on their feet. Physician Ron Sherman, Monarch Labs' founder, started out raising maggots in a closet at the VA hospital in Long Beach—a closet that "became quite spacious once everyone found out what I was doing." The moment a fly would get loose, the administration jumped on him. Sherman has since moved his "living medicine" operation to a warehouse near the Irvine airport, where he raises maggots, leeches, and fecal bacteria (for transplants). I can imagine the company's Schedule C taxable expense form simultaneously attracting and deflecting a visit from the IRS.

FILTH FLIES are lured by the odor of decay: a whole body or sometimes just a part. A moist, rank, infected body opening—be it a wound or a natural cavity—is a VACANCY sign to a gravid female. When maggot infestation shows up in a medical journal, it's generally accompanied by the technical term for it, "myiasis," and a revolting photographic close-up of the infected, infested part: gums, a nostril, genitals.

Here again, some words from the Armed Forces Pest Management Board: "Vaginal myiasis is a concern of increased importance because of the larger numbers of women serving in deployed units. . . . Egg laying may be stimulated by discharges from diseased genitals." In a hot climate, there might be a temptation to sleep outside uncovered, the board points out. And the kind of soldier who sleeps outside with no underpants would also, I suppose, be the kind of soldier with a genital disease. The kind headed for "dishonorable discharge" of one kind or another.

And finally, there is "accidental myiasis," typically of the intestines. The tale unfolds like this: The patient espies maggots in or near his daily evacuation and assumes he has shat them out. He further assumes—as does his doctor—that he accidentally ate some food infested with fly eggs. One hyperventilating MD, writing in a 1947 issue of *British Medical Journal*, claimed that the "resistant chitinous coating of the egg" survives the acids and enzymes of the stomach, enabling the larvae inside to travel unharmed to the less hostile environment of the intestines, where they would hatch and set up camp.

To the rescue, in the form of a letter to the editor, comes F. I. van Emden, of the Imperial Institute of Entomology. Does it not make more sense that the larvae were hatched not inside the patient but inside—as Van Emden put it, giving toilets and bedpans the ring of religious sacrament—"a vessel used for receiving . . . the excrements"? Furthermore, Van Emden points out, insect eggs are not made of chitin. The "shell" is a fine, thin, permeable membrane. To prove his point, Emden set up an experimental tabletop stomach, a mixture of warmed gastric juices and chewed bread, into which he placed eggs and larvae of the species in question. The larvae, including those inside eggs, were killed.

To any in need of further reassurance, I give you Michael Kenney, of Governmental Medical Services for the city of Katanga in the Belgian Congo, circa 1945. Presumably the GMS was an agency providing health care for indigents. "Sixty human volunteers . . . ," Kenney wrote, in *Proceedings of the Society for Experimental Biology and Medicine*, "were fed living maggots" of the common housefly, encased in large gelatin capsules. It's unclear whether the larvae—twenty per subject!—were encapsulated individually or inhabited one large community capsule, but either way it took two glasses of water to get them down. A third of the time, the capsules were vomited up shortly after they were swallowed, their passengers still for the most part alive. In the remaining two-thirds of the subjects, diarrhea with dead maggots ensued. An "occasional" maggot survived the odyssey, but that doesn't mean the volunteer was infested. A brief transit through the alimentary canal is different from settling in and passing your childhood there. All the volunteers' symptoms cleared up within forty-eight hours and no further maggots appeared. This suggested that, first, fly larvae "do not produce a true intestinal myiasis in man." And second, there's no such thing as free health care.

It's almost 8:00 p.m. at the Peck residence. George has brought out a tray of pinned insect specimens. I'm distracted at the moment by a live one.

"George?"

"Mm?"

"You have a large, somewhat frightening insect on your shoulder."

Peck doesn't bother to confirm this. Without removing his gaze from the tray, he says, "It's probably a brown marmorated stink bug." This time of year they're apparently everywhere. He explains that the name derives from the smell released when the bug is crushed. This one isn't crushed but carefully escorted out the screen door into the deepening Maryland dusk. Peck sits back down at the kitchen table. "They're beautiful under a microscope."

• • •

ETTING ASIDE George Peck—an act I've put off for as long as possible—most of the military's filth fly researchers are down in Florida. The Navy Entomology Center of Excellence (NECE) is located in Jacksonville, about an hour's drive from colleagues at the US Department of Agriculture Mosquito and Fly Research Unit. NECE serves as the military's pest control arm. It is a job that will go on forever. Because new generations come and go in a matter of weeks, flies quickly evolve resistance to whatever new pesticide they're hit with. There will always be some with a mutation that helps them survive, and those survivors and their rapidly proliferating spawn will repopulate the area, laughing at the humans with their misters and foggers and truck-mounted sprayers.

The flies of the Gulf wars are recalled as maddeningly persistent, a function of food's relative scarcity in the desert. During Operation Desert Shield, Navy entomologist Joe Conlon camped with a light infantry battalion in the Saudi Arabian desert near the Kuwait border. The flies served as an unpleasant but effective alarm clock. "You'd be asleep with your mouth open. Soon as dawn came the flies would be out, looking for food and moisture. They'd fly right in your mouth. You'd wake up to the sound of Marines coughing and cursing." USDA fly researcher Jerry Hogsette told me about a team of entomologists in Operation Desert Storm who drove off into the empty desert until they could no longer see the base, stopped, and opened a can of sardines. Within seconds, there were flies.

The fly's tenacious commitment to humans and their filth explains the military's enduring commitment to extermination: Soldiers constantly waving off flies are soldiers poorly focused on their job. When the job involves shooting and not getting shot, that's a hazardous distraction. With livestock, too, the distraction can be lethal. Hogsette says a cow can become so focused on shooing flies that it forgets about eating and starves. The agricultural community uses the term "fly worry."

The Gulf wars saw a related condition: insecticide sprayer worry. Shortly after the United States arrived in Kuwait, military intelligence determined that Saddam Hussein had purchased forty insecticide sprayers. With all the talk of "weapons of mass destruction," paranoia was running high. Joe Conlon was brought in to assess the likelihood—and the danger—of the devices' being used to disperse chemical or biological weapons. He deemed it unlikely. "You can't control where the cloud goes. You're just as likely to poison your own troops." Conlon's professional opinion was that Saddam Hussein wanted to kill some flies.

High-volume fly traps are a popular tool on military bases, because they're low-maintenance. Here the artistry is in the lure. NECE has tested different wavelengths of ultraviolet light, varied background colors, and all manner of chemical attractants. There was a fleeting moment, during World War II, when fly attractants played a more strategic battlefield role. Nazis had poured into a Spanish enclave of Morocco with the aim of cutting off the Allied supply line to troops fighting Erwin Rommel's Afrika Korps. The Pentagon called upon Stanley Lovell, director of research and development for the Office of Strategic Services or OSS (precursor to today's CIA), to devise a way to quietly, as Lovell put it in his memoir, "take out Spanish Morocco."

"I evolved a simulated goat dung," Lovell wrote, improbably. Spanish Morocco being a land with "more goats than people," the decoy dung would, he reasoned, fail to arouse suspicion. The plan was to spike the turds with both a powerful fly attractant and a cocktail of pestilent microorganisms and then drop them from planes during the night. Filth flies would take over from there: landing on the dung, picking up pathogens, and delivering their deadly payload to the Nazis' meals.

The OSS files in the National Archives and Records Administration include dozens of entries for gadgets and weapons dreamed up by staff,* but I found nothing under "goats," "dung," or Lovell's name for the project, Operation Capricious. Lovell wrote that he and his colleagues were "well along" with it when word arrived that the Germans had withdrawn from Spanish Morocco. Perhaps. I suspected that the killing shit never made it further than the drawing board. Or, more likely, the cocktail napkin.

And then I came upon an OSS file labeled "Who, Me?" And it was clear I had underestimated Stanley Lovell.

---

* My favorites, in alphabetical order: ashless paper, boosters and bursters, collapsible motorcycles, Hedy Lamarr, luminous tape, nonrattle paper, paper pipes, pocket incendiaries, punk type cigarette lighters, smatchets, sympathetic fuses, and tree climbers.

10 What Doesn't Kill You Will Make You Reek

A brief history of stink bombs

THEY WERE MY FIRST secret documents, and they did not disappoint. Individual pages in the file were rubberstamped "SECRET" in oversize letters, once in the top margin and again at the bottom. An additional, wordier rubber stamp warned that the document contained "information affecting the national defense of the United States of America," and that transmission of its contents was a violation of the Espionage Act. Some of the papers were marked for delivery "by safe hand," the hand belonging to an ambitiously vetted government courier whose fine leather satchels no Customs agent was allowed to inspect.

The first document in the file was dated August 4, 1943, and addressed to Stanley Lovell at OSS—America's espionage agency during World War II. It was from one of his British liaisons, and it referenced a memo from a gadget man in the British intelligence agency SOE (Special Operations, Executive). "Replying to your letter of the 6th July," the memo began, as though we were headed for some dusty government protocol. And then veered abruptly off the tracks.

"Up to the present our employment of evil-smelling substances has been mainly for the purpose of contaminating individuals' clothing." The memo contained the secret formula for "S liquid" (S for stench), an oily mixture with a "highly persistent smell suggestive of personal uncleanliness." Included as well were plans for two delivery systems: a gelatin capsule to be pierced with a pin, squirted, and "dropped immediately once the operation is completed," and a small brass spritzing device midway between a perfume atomizer and a pesticide sprayer, the latter to be "carried about secreted in the hand or pocket." I pictured a British operative's wife stumbling upon one in the pocket of her husband's blazer, lifting it in front of her face and squeezing the bulb, expecting cologne and receiving instead another disquieting hint that her spouse was not wholly forthcoming about the particulars of his career.

The goal: "derision or contempt." The demoralization and alienation of German and Japanese officers in occupied countries. Allied espionage agencies were keen to find cheap, low-profile devices to put into the hands of saboteurs and resistance groups, motivated civilians eager to help the cause.

The OSS, with help from weapons developers at the National Defense Research Committee (NDRC), got to work on a stench of their own devising. In his memoir, which was published twenty years after the war, Lovell downplayed SAC-23—"Contaminator, Stench"—as "comic relief," an antidote to the "grim, bloody and sordid" job of creating "new and special weapons to kill people." But the thickness of the files and the detailed, deadpan formality of their contents indicate otherwise. The SAC-23 project dragged on for two full years, sapping the patience and clogging the in-boxes of seven majors,* eight lieutenants, four captains, and a wing commander.

Lovell's original directive, as stated in his memoir, was for a substance with the "revolting odor of a very loose bowel movement." ("Who, Me?" was Lovell's cover name for SAC-23.) He wanted something to distribute among Chinese resistance elements for the purpose of humiliating Japanese officers. For some reason, Lovell believed the Japanese to be uniquely vulnerable to harassment of this nature: "A Japanese thought nothing of urinating in public, but he held defecation to be a very secret, shameful thing." (Like racism.) The NDRC set forth additional requirements. It should have a "range" of at least ten feet "without backfire." "It should be silent in operation." Inconspicuous. Impervious to rain, soap, solvents. Conferring shame for a minimum of several hours.

A chemical engineering firm out of Cambridge, Massachusetts, was brought on board to develop the formula. The Arthur D. Little Company put their top odor and flavor man—their "Million Dollar Nose"—on the job. Ernest Crocker rose to the challenge like the reek off a landfill on a summer's noon. "A stink among odors may be compared to a weed among plants, . . . a plant out of place, such as a potato in a flower garden," he wrote in a background memorandum. In other words, context is key.† At an Italian deli counter,

---

* Including—attention, aging M*A*S*H fans—a Major Frank Burns. I nursed a fleeting hope that a Major Houlihan would appear on a CC list alongside him, but it was not to be.

† For years, the most-requested scent at eclectic fragrance firm Demeter was newborn baby's head. So they isolated and synthesized it. (Weird? Not for Demeter. Their line also includes Laundromat, Mildew, Paint, Play-Doh, Dirt, and Pruning Shears.) Baby's Head did not test well. Outside the context of a baby, it turns out, newborn scalp odor isn't well liked. The firm added baby powder and citrus notes and changed the name to New Baby. Baby's Head perhaps making some people uncomfortable, what with Pruning Shears right next door.

a whiff of butyric acid reads as parmesan cheese; elsewhere, vomit. Likewise, the odor of trimethylamine can be described as fishy or vaginal—as Crocker coyly put it, "pleasant or unpleasant according to circumstances." Very few smells can be classed by their very nature, regardless of context, as repulsive. These were the ones the OSS needed: the "stink-makers."

The main active ingredient of the Brits' "S liquid" was skatole, an intensely fecal-smelling* compound produced by gut bacteria as they break down meat. Thus wrote Crocker in a second memo, breezily entitled "Facts about Feces." Acids from digested carbohydrates, he went on, yield the sour notes of flatus. Trace amounts of hydrogen sulfide provide the telltale rotten egg stench. And so on. The takeaway being that it's no simple task to make a man smell like shit. The smell of human feces is, like any in nature, staggeringly complex, comprising dozens if not hundreds of chemical compounds. (This is why novelty "fart sprays" smell abominable but not especially like farts.) A high-fidelity simulant would be a daunting and costly prospect.

And, Crocker felt, not optimally effective. "In general," he wrote, "a mixture is best for it bewilders." Olfaction, like taste, is a sensory security guard, an early detection system for chemicals that might be harmful. If you don't recognize the smell, you can't know that it's safe. Over the millennia, humans who played it safe and backed away from strange smells were more likely to survive to pass on their genes. Thus an unidentifiable foul smell is a more potent weapon than one that is merely foul.

The more typical military application for "malodorants"—as

---

* In highly dilute form, skatole adds a flowery note to perfumes and artificial raspberry and vanilla flavors. This I learned from HMDB, the Human Metabolome Database, which I consult the way normal people consult IMDb.

modern olfactory nonlethal weapons are known—has been "terrain denial": keeping (or getting) people out of a targeted piece of real estate: a Viet Cong tunnel, a terrorist hide, a weapons cache. Almost always, these have been cocktails of odors. In its "evil-smelling substances" memo, the SOE described tiny glass stench ampules distributed to resistance groups to be dropped on the carpet in known Nazi meeting places. The officers would unknowingly crush them with their showy black lion-tamer boots, and an unplaceable, rank sulfur-plus-ammonia stench would fill—and then empty—the room.

Our man Crocker got to work. He set up in-house "organoleptic testing" sessions to gauge the effectiveness of dozens of heinous blends. *Organoleptic* means "involving the sense organs." It means you didn't want to be employed at Arthur D. Little in the later months of 1943. Crocker eventually settled on a blend of skatole; butyric, valeric, and caproic acids; and a mercaptan: shit, vomit, smelly feet, goat, and rotten egg. Samples were prepared and delivered to the NDRC in two formats: a more intense "paste-form stink," for smearing, and a liquid stink in a squirtable two-ounce lead tube. Crocker assured his clients that the latter would render a target "highly objectionable for not less than two hours at 70 F." He promised nothing short of "complete ostracism," concluding his report with a tagline surely unique in the annals of marketing: "as lastingly disagreeable as a product of this kind can be."

P AM DALTON has a bottle of Who, Me? in her lab. Dalton works at the Monell Chemical Senses Center, an independent nonprofit with ties to the nearby University of Pennsylvania and a long history of Defense Department–funded malodor research and a four-foot-tall bronze nose out in front. I first met her in 1997, when she was serving as an expert witness in a pig farm lawsuit. She was the saucy redhead happily walking the fence lines, sampling fumes with a handheld electronic nose. She has a few more laugh lines now, but her hair is still red and she still loves her work.

The Dalton Lab has no detectable smell, though there are many things in here that stink. On a shelf over our heads is a box of firefighter underarm odor, each subject's contribution sealed in a Ziploc bag. Where other refrigerators would hold ketchup and salad dressing, Dalton's has a bottle of civetone, a synthetic version of the anal scent gland secretions of the civet cat. The weapons-grade smells are over beneath the fume hood. The Who, Me?, which tends to cause panic if it gets into the building's ventilation system, has been bottled, taped, double-bagged, and entombed in a small, reclosable can.

Dalton Lab manager Christopher Mauté unwraps it for me. Mauté has high cheekbones, an aquiline nose, and glossy dark hair that sweeps back from his romantic-lead hairline seemingly without the help of product. Underneath he is all science. He is, by his own description, the guy who smells the roses at the wedding reception and says, "Mmm, phenylethyl alcohol." Mauté holds the opened bottle near my face, although doesn't relinquish his grip on it. If I drop the Who, Me? all of Monell West becomes, as Ernest Crocker liked to say, highly objectionable.

A tentative sniff confirms that it is awful, but it's not what I had expected. This particular version landed far afield of Lovell's original diarrheal objective. Dalton just returned from the Milpitas, California, dump, and the smell takes her straight back there. It's sulfury, but not in a jokey-farty rotten egg way. It's got a meaner, spikier disposition.

Mauté recaps the Who, Me? and reaches under the hood for

another bottle. A hand-printed label says Stench Soup and, in larger letters, DO NOT OPEN. In 1998, the Joint Nonlethal Weapons Directorate commissioned Monell to develop a superlative malodorant for clearing buildings or benignly dispersing a violent mob. Stench Soup is what Dalton's team came up with.

Mauté holds the cap in front of my face. Dalton takes two steps backward, anticipating the moment when the smell climbs from its hole. It's bad, gag-bad. It's Satan on a throne of rotting onions. Mauté quickly closes it and puts the container back.

"Put the hood down," Dalton says, calmly but firmly. And then, less calmly: "PUT THE HOOD DOWN."

It seems like a good time to go out and get some lunch. I follow the pair to a nearby oyster house to hear the story of how the world's most objectionable smell, the mixture known as Stench Soup, came to be.

It began with a confection called US Government Standard Bathroom Malodor. (The government being the developer of the smell, not its natural source.) The smell was developed during World War II as part of an effort to create a compound for deodorizing field latrines. I have seen old photographs of these: the row of grinning GIs, naked rear ends slung over a log fence rail erected above a pit. To test the various deodorizers they concocted, Army chemists needed to re-create the stench in the lab. It was apparently unique. "Open field latrines used by hundreds of men over an extended time, often in sweltering heat, don't resemble very well what occurs in a typical residential bathroom," says Michael Calandra, of the flavor and fragrance company Firmenich. Firmenich has an entire "library" of malodors available to industry for testing cleaning products and deodorizers.

"So we took Bathroom Malodor," Dalton says, forking something trimethylaminey, "and we sweetened it up a little." The inspiration was provided by a panicked Las Vegas hotel owner who had telephoned Dalton after his sewage pipes backed up. The addition of the flowery-smelling cleaning product he'd used had made the smell yet more odious.

Mauté reveals the other reason Stench Soup includes a fruity top note. "Most people when they do an initial sniff, it's shallow. And then if the top note is pleasant, they're willing to embrace it."

Dalton jumps in. "So the sweet note hits you, and you inhale more deeply and—" They're like excited siblings home from a field trip.

"—and that sulfur is just waiting for you on the keeper inhale."

"And the sulfurs, once they get in your nose? They last. They get trapped in the mucous. They keep rebinding to the same receptors."

It's impressive, the ingenuity that goes into a top-flight noxious-smelling substance. The British S liquid included a compound that delayed the onset of the odor, thereby "improving the chances of the operator being able to escape before the smell was detected."

Would that wars could be fought and won this way—with weapons that didn't kill or harm. If sacrificing lives for the larger good of nation or cause were not part of the moral equation, imagine the enterprise that would have gone into morale-sapping instead of atom-splitting and armor-piercing. In the same delightful category as Stench Soup, we have the brainchild of Bob Crane, a materials engineer at a research lab at Wright-Patterson Air Force Base who attended a nonlethal weapons brainstorming session during Operation Desert Storm.

Crane set the scene for his idea. The enemy is hunkered down, taking fire. Days go by. The supply lines are cut off. The men are hungry, lonely, angry. Now you introduce the secret weapon: the nostalgic aroma of fresh-baked bread. Crane is an expert in micro-encapsulation, the technology behind, among many other things, scratch 'n' sniff. It's possible to encapsulate a scent in tiny grains of a powder that could then be dropped over the enemy position while the fighters sleep. The next day they walk over the microcapsules, breaking them open and releasing the scent. It's too much. They miss home, they miss their mother, they decide to desert.

A S CROCKER promised, SAC-23 stank "lastingly." No one knew this better than the quality control testers of Maryland Research Laboratories, to which the OSS had shipped off a box of two-inch lead tubes of it. "Almost without exception," states the report, "the operator was contaminated when squirting the contents of the tube."

A Major John Jeffries of the OSS ran some tests of his own. Twelve percent of the tubes, he wrote in an acid letter from July of 1944, were leaking *when they arrived in his office*. When ten of the nonleaking tubes were placed in an oven to approximate warm weather warehouse conditions, all commenced to ooze. To assess the real-life practicalities of discharging SAC-23, Jeffries dressed a dummy in a military uniform. One out of three tubes "backfired" onto his hand. Even just unscrewing the cap, he wrote, it was "impossible to prevent getting some of the liquid on my hand."

Storage and dispersal issues have continued to vex the malodorant community. There were so many problems with a Vietnam-era concoction that the developers considered a binary delivery system, two compounds kept segregated like epoxy components, producing a stench only in combination. Dalton told me the story of a catastrophic misfire during a test of Stench Soup. To contain the stink, Dalton had subjects wear an airtight plastic hood. "It was like a biosafety suit, only we were putting the contaminated environment *inside*." A flexible tube delivered malodorized air through a sealed port in the hood. On the third day, the system malfunctioned. Instead of pumping in Stench Soup as a gas of carefully calibrated concentration, it bubbled in the undiluted source. The subject happened to be one of Dalton's military funders. When it was over, the man pulled off his hood and reached up to find the whole back of his head saturated with the oil. Dalton was speechless. "I kept opening my mouth like a carp, and nothing would come out. My technician goes, 'Gosh, was it on you when you got here?' Trying to put the blame on him! Like, 'Perhaps your hair gel?'" The man was heading straight from Monell to the airport. "We had to take him up to the animal floor and let him take a shower."

Meanwhile, back in time, the OSS had a bigger problem. The tubes were defective, and there wasn't time to redesign them. Someone had gone ahead and added Who, Me? to the OSS catalog. Urgent cable requests were pouring in. Ten thousand tubes on order. "Memorandum on 'Who, Me?': Prevention of Contamination of the Operator" details the agency's scramble for a fix. Slip-on paper hand shields? Too flimsy. Cloth-backed paper shields proved sturdier, but conferred protection only "when one squirts from a horizontal position."

In the end, the OSS opted for rubber shields, despite a domestic rubber shortage dire enough to have prompted tire rationing and soon-to-be collectible posters (America Needs Your SCRAP RUBBER). Along with gas masks, life rafts, and jeep tires, the nation's wartime rubber needs would come to include rubber Who, Me? sleeves with anti-dribble operator-protection lip.

Late in 1944, 95 rubber-accessorized Who, Me? tubes were rushed to Maryland Research Laboratories. They passed the Rough Handling Test. The Accelerated Aging Test. The Tropical Weathering

Test and the Arctic Storage Test. The Combined Rough Handling and Tropical Weathering Test. Only once was a tester's hand contaminated, owing to "a strong wind blowing across the direction of fire." At last! The report of this final round of testing, dated November 9, 1944, pronounced Who, Me? ready for production and shipment to the field. Federal Laboratories was persuaded to take the order: 9,000 units, at 62 ½ cents each—enough revenue to cover the purchase and installation of the very finest fume hoods money could buy.

And there the story should have ended. But did not. Ernest Crocker, sensing the trough of lucrative government contracts being pulled from under his snout, dropped a stink bomb of his own. "The odor of Who, Me? is not considered objectionable by Orientals." Humiliating the Japanese in occupied China, you will recall, had been Stanley Lovell's original goal. Crocker offered to develop a new malodorant. Production was delayed yet again. More tests ordered. Your taxpayer dollars shaking their little green heads in disbelief.

"In discussions with a Navy physician who had dealt a great deal with Oriental peoples," reads the Arthur D. Little company's February 19, 1945, Supplement to Final Report on Who, Me?, "the conclusion was reached that only two types of foulness could be counted upon as certainly objectionable: skunky odors and cadaverous odors. "With 'Who, Me?' as a pattern, but with skunkiness substituted for fecal odor, we produced 'Who, Me?– II.' This preparation has an atrocious odor, with pronounced penetrative and lasting qualities. It is reasonably certain that it will fill all Japanese requirements." Five hundred Who, Me? and one hundred Mark II Oriental Who, Me? tubes were finally manufactured.

Not a single one was shipped to the field. Why? Because the National Defense Research Committee had been working on a far more lasting and penetrative weapon for use against the Japanese. Seventeen days before the second and *final* Final Report on Who, Me? was released, the United States dropped an atomic bomb on Hiroshima.

O N A fifteen-hour flight, it is not unusual to notice an unpleasant bathroom odor or even, depending on how much turbulence there's been, the smell of vomit. It is unusual, though, to notice these smells emanating from an overhead compartment. Six hours into a flight to South Africa, that is what began to happen to Pam Dalton. "I had stood up for the first time, to go to the restroom, so my nose was right about at that height. I thought, *Holy shit, those are my odors.*"

The year was 1998. Dalton was undertaking research for the US military, still hot in pursuit of the Holy Grail of malodors, the "universally condemned smell." She was traveling to Africa on an unrelated project and had decided to bring an assortment of malodors to test on Xhosa residents in a nearby township: one more culture weighing in. In her carry-on were bottles labeled Vomit, Sewage, Burned Hair, and US Government Standard Bathroom Malodor. Dalton had sealed and double-bagged the bottles but failed to take into account the change in cabin pressure. The liquids had expanded and leaked around the edges of their paraffin seals. Fortunately the overhead bin contained only her and her companion's luggage. "I told him, 'You can't get anything out of the overhead compartment. Everything here at our feet is all we can have for the entire flight.'" As long as the compartment stayed closed, the odors would be largely contained—until the plane landed. And then what? "Here was my cleverness. I didn't open the compartment until they had opened the doors of the plane. I figured that way people could blame it on something coming in from outside."

Prior to her work with the Xhosa subjects, Dalton had exposed Asians, Hispanics, African Americans, and Caucasians to these same smells. The stand-out winner? US Government Standard Bathroom Malodor. "People hated it. They really, really hated it, and they thought it was dangerous." Ernest Crocker was wrong about the Japanese. Among Dalton's Asian subjects, comprising Japanese, Korean, Chinese, and Taiwanese, 88 percent—the highest percentage among all the ethnic categories—described it as an odor that would make them "feel bad." It took the top slot in an Odor Repellency Ranking among all five ethnic categories. It repelled, by and large, everyone, with the exception of one unusually open-minded individual who judged US Government Standard Bathroom Malodor to be a "wearable" scent.

None of Dalton's other bottled vilenesses approached a workable criterion of universality. Sewage Odor was no good at all. Fourteen percent of Hispanic subjects described it as an odor that would make them feel good. Around 20 percent of Caucasians, Asians, and black South Africans thought it smelled edible. Vomit Odor made a similarly poor showing, with 27 percent of Xhosa subjects describing it as a feel-good smell and 3 percent of Caucasians being willing to wear it as a scent.*

Dalton's colleague Gary Beauchamp, Monell's director at the time I visited, had had high hopes for Burned Hair, a stand-in for

---

*This is not the reason International Flavors and Fragrances developed a proprietary vomit scent. They did so at the request of a company that planned to market it as a diet aid, a stick-up odor dispenser that would discourage you from eating by making your refrigerator stink like vomit. The item was never produced, because in tests, a certain percentage of people, particularly if they were hungry, had a positive response to the smell. They wanted to have it as a snack.

burning human flesh—an odor he felt confident all cultures would detest. What atrocities had Beauchamp spent time downwind of that he would have had this insight? "Nothing like that," said Dalton. "He told me he used to peel skin off his fingers and put it on his coworkers' lightbulbs, as a practical joke." When the bulb was turned on, the heated peelings would start to smell. "I said, 'Well *that's* a side of you I didn't know.'"

Unlike Vomit and US Government Standard Bathroom—malodors that exist ready-made—burned flesh/hair odor had to be concocted from scratch. Dalton convinced her hairdresser to collect a bag of floor sweepings, which she brought to the lab and pyrolized— pyrolization being a science lab version of leaving it on someone's lightbulb. A mineral oil was infused with the collected vapors, and this is what subjects sniffed. Forty-two percent of Dalton's Caucasian subjects thought Burned Hair smelled edible. Six percent of the Xhosa would wear it as a scent.

No one, it seems, wants to eat, wear, or be anywhere near the smell of a military field latrine. And so it was that Standard Bathroom Malodor became the starting point for Stench Soup. How has the smell served its country in the years since? Dalton shrugs. "I gave them the recipe. I have no idea what they did with it."

F YOU ever visit the Monell Center, chances are good you'll be pressed into service as an "odor donor." Someone will want to collect your breath or sniff your earwax or gather the gases exuding from your underarms. Chances are also decent that the study for which you have donated your aroma is funded by the United States Department of Defense. Of late, the military has taken an interest in the smell of stress. Were there a signature scent consistent from one stressed person to the next, something a sensor could pick out amid the clamor of perfume and cigarette smoke and last night's garlic fries, then a sort of BO profiling could be undertaken. Sensors could be set up in airport security to identify suspected terrorists—though care would have to be taken to distinguish bombers from nervous fliers.

Body odors might also be used to monitor the stress level of personnel in high-pressure, high-risk jobs. A chemical sensor could be part of a so-called smart uniform. If stress compounds could be reliably detected on breath as well, the sensor could be part of a helmet mouthpiece. "We're doing a pilot study for the Air Force," Mauté told me. A pilot pilot study.

The goal would be intervention. If you hit a level of stress likely to compromise your ability to complete a task safely, an alert could be sent wirelessly to superiors. Your BO quietly turning you in. Alternatively, some kind of automatic intervention—say, a cutoff of the equipment—could be triggered.

I donated some stress smell earlier. Mauté had me put gauze pads in my armpits and count backwards by 13 from 200 while he timed me. When I made a mistake, I had to start over. At one point he threatened to post the footage on YouTube. My armpit gauze has been tweezered into a glass specimen jar like an exotic lace-winged insect. Mauté, having sniffed it, pronounced it "a wonderfully fresh BO smell."[*] Every now and then in life, a compliment is tucked so seamlessly into a insult that it's impossible to know how to react. Around Monell, body odor seems to confer no shame. Seems to

---

[*] As opposed to a "stale-uriney BO smell," the smell my stepdaughter Phoebe, as a little girl growing up in a big city, called hobo pee. Monell Chemical Senses Center BO expert Chris Mauté surmises that "hobo pee" is the smell of sweat and sebum that has been extensively broken down by bacteria: "the kimchee of body odor."

possibly even confer respect. When Mauté referred to a colleague as "*the* donor in terms of his ability to produce body odor," it seemed a kind of honorific, so much so that only much later did it occur to me to omit the man's name.

The odor descriptor Monellians use for human flop sweat, the emotion-moderated secretions of the underarm apocrine glands, is "onion-garlic-hoagie." Presumably there are odor descriptors for the smell of other animals' stress, but you'd have to ask those animals, or the animals who hunt or harass them. If you wanted to know what distressed groupers smell like, for example, you could ask a shark. Or you could ask the US Navy.

11 Old Chum

How to make and test shark repellent

IF YOU WERE IN the market for a chemical that is harmless to humans but toxic to lesser classes of creature, you might reach out to someone in agriculture. A good pesticide, if there can be such a thing, combines both qualities. The insecticide rotenone was the topic of a 1942 memo from the US Department of Agriculture to the US Eleventh Naval District. In addition to killing bugs, rotenone, the letter stated, is a powerful fish poison. When added to water at concentrations only feebly toxic to humans, the chemical "stupefied goldfish."

This was encouraging information, except that the Navy had inquired about something for sharks. World War II marked the first time in US military history that battles were being fought on and over tropical seas, and stories had begun to circulate of sailors and fliers being attacked and devoured after abandoning ship or ditching their plane. (During the previous world war, crews wound up in the North Atlantic, where cold devoured them first.) One particular narrative made its way to a man named Henry Field (as in the Field Museum of Natural History), who at the time held the title

Anthropologist to the President, as well as a post with—well, hello again!—the OSS.

In June 1941, the story went, an Ecuadorian Navy plane went down in the Pacific after running out of fuel. The "desperation and terrification" of the flight officer is detailed in the official report of the incident, which Henry Field either heard about or read. It was a moonlit night. The man wore a life jacket, and as he swam he pushed along the body of a drowned colonel. Sharks began to cross the water in front of him. "At a given moment I felt that they were trying to take away the corpse, pulling it by the feet, on account of which I clutched desperately the body of my companion and together with him we slid until the tension disappeared." Here I confess I became more interested in the translator of the report than its terrificated protagonist: "Once refloated, with despair I touched his legs and became aware that a part of them was lacking." The flight officer abandoned the demi-corpse and continued alone to shore, "with various sharks following."

"Night after night," Henry Field recalls in his memoir, "I thought of these men . . . with sharks cutting through the water around them." As Anthropologist to the President, he had Franklin Delano Roosevelt's ear, even, it seemed, in matters of ichthyology. "I wrote the president a memorandum suggesting that we try to develop a shark repellent."

With presidential blessings, Field met with fellow museum curator Harold J. Coolidge, also on the payroll of the OSS. Coolidge was a primatologist—a silverback gorilla he collected (shot) in the Congo resides to this day in Harvard University's Museum of Comparative Zoology—but he agreed to oversee the shark project. You can well imagine that a gorilla expert on salary with a spy organization might suffer a mild sense of purposelessness. Here at last was something up his alley, if not precisely on his doorstep. Coolidge hired another curator pal, W. Douglas Burden, as the project's principal investigator. Burden was an expert on Komodo dragons, had written an entire book about Komodo dragons, but he, too, knew little about sharks.

For actual shark expertise, the OSS turned to a college dropout named Stewart Springer, whose résumé included stints as a commercial fisherman and as a chemical technician at the Indianapolis Activated Sludge Plant. In 1942, there were no experts in shark biology and behavior. Truly, no one knew much about the creatures. The combination of hands-on shark experience and sludge chemistry was, in fact, ideal background for the task. "Dr." Springer, as some of the OSS correspondence refers to him, was as good as it got.

The US Navy agreed to contribute funding, even though, as one of their rank pointed out, there was at that time no formal record of anyone who had taken the oath of the Navy having been harmed by sharks. Their concern was morale. Fear of sharks, however irrational, was thinning the ranks of willing fliers. Stewart Springer voiced the cockamamie irony of it: "It was okay to give one's life for one's country, but to get eaten for it was another matter." If nothing else, a repellent would serve as what Douglas Burden called a "pink pill," a psychological fix for shark-shy aviators. On July 3, 1942, funding was approved for OSS Office of Scientific Research and Development Project 374, Contract OEMcmr-184: a three-month investigation "looking to the development of means of protection against sharks, barracuda and jellyfish* for men adrift in lifebelts." (In three hun-

---

* Two months into it, the Chief of the American Intelligence Command wrote to Harold Coolidge urging him to add piranhas to the list. AIC needed better piranha intelligence. Years ago, nature filmmaker Wolfgang Bayer told me the story of the time he was sent to the Amazon to get footage of bloodthirsty piranhas devouring a capybara. Bayer strung nets across the river to trap a school of piranhas. He captured a capybara and herded it into the river. Nothing. He starved the piranhas. Still nothing. He went home.

dred some pages of archived correspondence for Project 374, I saw but two passing references to barracudas. As far as I can tell, no one ever got around to jellyfish.)

The lab work was done mainly at Woods Hole Oceanographic Institution, which housed a collection of captive sharks called dogfish—in size and temperament, somewhere between a great white and a goldfish. Rotenone was among the first substances the team tested. "Definitively negative," Burden reported to Coolidge. "Lethal doses do not deter the feeding process." The shark would die, but not before you did. Until such time as goldfish presented a threat to national security, rotenone would be limited to the arsenal of the USDA.

Seventy-nine substances were tested and rejected. Irritants failed. "Repulsive odors" failed. As did clove oil, vanillin, pine oil, creosote, nicotine. They tried compounds related to mothballs, asparagus, horse piss. The sharks ignored all of it. The first "hot lead" sprang from an item of sharker lore. Springer had heard that a shark carcass abandoned on a bait line will ruin the spot for shark fishing. He and his team got to work. They rented an "isolated" house in Florida for $10 a month, and never, I'm guessing, was a cleaning deposit more roundly withheld. Chunks of shark muscle tissue were left out at room temperature for four or five days. An extract was then prepared by grinding the decomposed flesh, stirring in alcohol, and filtering the resultant shark muck.

Forty-three experiments later, Springer enthused in a note to Burden, it was possible "to say POSITIVELY that the meat contains some substance strongly repellent to sharks." Repellence value 88.4 percent! Ninety to 100 percent effective! The bimonthly progress report of Contract OEMcmr-184 describes Springer as "sufficiently convinced of the effectiveness of the concentrate that he would be willing to test it in a life belt with a bucket of blood."

An expedition to test the decayed meat concentrate on wild sharks had been slated as the next step, but Springer and Burden urged the OSS to begin production immediately. "If we really have something now and . . . the field test delays use of a good thing by six months," Springer wrote to Coolidge, "and if during those six months . . . some poor devils might have been protected it would be bad." Springer happened to know a contractor who could get right to work producing the concentrate. Shark Industries was a Florida purveyor of shark skins and shark oil—and also, speaking of things that smell fishy, Springer's sometime employer. The company, Springer felt certain, would be able to produce enough shark extract to outfit 2,000 to 5,000 life jackets per month. If Springer had his way, the whole undertaking would soon be moot, as there would be no sharks left to repel.

The OSS didn't bite. Rather than move forward with the concentrate, they wanted to try to isolate the active ingredient—a compound that could be ordered or cheaply synthesized, thereby saving them the cost and bother of large-scale shark carcass reduction. Chemists were hired, three of them, and they soon came up with a promising candidate: ammonium acetate. It, along with two compounds that had earlier shown promise (copper sulfate and maleic acid), plus thirty pounds of the Macbethian-sounding "extract of decomposing shark meat," were flown down to Ecuador, to the very same waters where our story began, to be tested on "voracious surface-feeding sharks." Lodgings were secured, boats and guides hired. Three weeks later, Burden dispatched a glum cable: "The waters off the coast of Ecuador have been virtually empty."

From deep in the pockets of the OSS came Harold Coolidge's reply: Try Peru. "Don't be discouraged," he wrote. "Shark hunting is not unlike tiger hunting. You remember how plentiful tigers are in various parts of French Indo-China until you reach the point when you want to shoot one and have only two or three weeks at your disposal." You got the sense, leafing through these letters, that a career in natural history was little more than a way for well-connected gentlemen to finance far-flung safaris and fishing expeditions in the name of science. The title of Douglas Burden's memoir nicely summed the job: *Hunting in Many Lands*.

The expedition eventually located some sharks, off the coast of Guayaquil, Ecuador. More discouraging words followed. Nothing worked. They tried combining the ammonium acetate and the copper sulfate, and that compound (copper acetate) seemed effective. Unfortunately, two or three pounds of it, in the form of a slowly dissolving cake (think urinal, not birthday), would be needed for one day's protection. This would not do. The Navy wanted something small enough and light enough—six ounces at most—to seal in a packet and attach to a life belt. The life belt, a precursor to the flotation vest, was a deflated rubber tube worn around the waist at all times and inflated in an emergency. Like any part of a serviceman's uniform, the belts developed holes from wear and tear. The last thing a seaman needed on top of a leaky life belt was a three-pound anchor of questionably effective shark repellent.

The Navy was losing patience. A hundred thousand dollars—$1.5 million in today's currency—had been spent, and they were no closer to having a practical, effective shark repellent than they'd been a year ago. The OSS was edged out, and the project taken over by the Office of Naval Research and the Naval Research Laboratory

(NRL). The first thing the Navy did was to make the field tests more realistic. Springer and Burden had been baiting lone meandering specimens—"casual sharks"—using hunks of mullet as their man-in-life-belt stand-ins. The NRL wanted a better approximation of the thrashing aftermath of a downed ship or plane and the "large schools of frenzied sharks" that that scenario was thought to attract and inspire. The so-called feeding frenzy was a state of mind in which, it was speculated, olfaction took a back seat to the "mob impulse." In August 1943, copper acetate was brought on board a shrimp trawler off Biloxi, Mississippi, and tested for its ability to protect "trash fish"—flailing, panicked specimens tossed off the back because they weren't shrimp. Guess what? Even *five to six pounds* of copper acetate per bushel of trash fish "did not by any means" interrupt the het-up mob trailing the boat. "The sharks hardly paused."

The final slap in the face of Project 374 would come in the form of a paper by Navy Captain H. David Baldridge Jr.: "Analytic Indication of the Impracticability of Incapacitating an Attacking Shark by Exposure to Waterborne Drugs." By plotting the speed of a closing shark against the speed of dilution and the concentration needed to put the creature out of commission, Baldridge showed that such a large quantity of drug would be needed that it "does not appear to be at all reasonable as an approach to the control of predaceous shark behavior." As one of Burden's colleagues put it: "You can't do much with a pint of liquid in an ocean."

Taking a cue from the octopus, Navy researchers next looked into using clouds of inky dye as a way to hide crewmen from potential predators. Under those same "mob psychology" conditions, all feeding activity was stopped until the dye had diluted to the point at which it no longer obscured the prey. Production began at once. Shark Chaser's active ingredients: 80 percent black dye and 20 percent pink pill—a little copper acetate having been added to the pot[*] for some false peace of mind. From 1945 all the way through to the Vietnam War, packets were available for the emergency survival supplies of lifeboats, life rafts, and life jackets on military vessels and planes. Even the post-splashdown survival kits of the Mercury astronauts were stocked with Shark Chaser.

Through all of it, there'd been skeptics among the Navy brass. Rear Admiral Ross T. McIntire, Chief of the Navy's Bureau of Medicine and Surgery, made the eminently reasonable point that a package labeled SHARK CHASER in bold capital letters might in fact lower, not raise, morale, planting, as it would, the seed of terror in minds that had been, until that moment, occupied by the real threats of ocean survival: dehydration, starvation, drowning, heat, cold. Especially given the "negligible danger," to use McIntire's words, that sharks posed to Navy personnel.

How negligible? Opinions varied, but at one point in the proceedings, the Commander of the South Pacific Fleet issued a memo to all naval bases and hospital ships soliciting "authentic cases of injury to personnel from attack by sharks." With all hands reporting, the final count was two cases. (One additional attack was later

---

* You may have heard stories about how Julia Child's first recipe was for shark repellent. Her OSS employment file shows that she indeed worked for the head of the shark repellent project, Harold Coolidge, in the Office of Emergency Rescue Equipment in 1944. However, her title was Senior Clerk, and her name appears nowhere in the OSS shark files. Child herself made no claim to have come up with the recipe for Shark Chaser but said merely that she followed it, mixing the ingredients "in a bathtub." This seems odd, as none of the other repellent prototypes were produced or tested at OSS headquarters. Leading me to wonder: Did she cook up Shark Chaser, or just a good story?

determined to have been a "vicious eel.") The OSS responded in time-honored intelligence-agency style: They disappeared the report. "The report on shark attacks has been destroyed, as you requested," reads an interoffice memo to Harold Coolidge from a staffer in December 1943.

It was another stink bomb for the OSS. They'd set out to develop a shark repellent based on one man's experience and another's political connections, with no solid data to support a need. If you look back at the Ecuador incident—the original impetus for all this—it really wasn't a testament to the danger or ferocity of sharks. If anything, it was a testament to the disinterest and/or shyness of sharks. The flight officer was adrift in a life jacket for thirty-one hours, yet he emerged from the ocean unmauled by the retinue of sharks that followed him most of the way to shore.

If you wanted to preserve morale, the better approach would have been to share these reassuring facts and statistics. "Correct information," wrote McIntire, "would be more universally operative in alleviating those fears than any repellent that could be devised." Beginning in 1944, that is what the Navy did. Their Aviation Training Division distributed copies of a pamphlet called *Shark Sense* to all future fliers: 22 pages of comforting facts, illustrated with comic drawings of cringing, perspiring, fleeing ("HALP!") sharks.

And it proved true. In a review of 2,500 aviators' accounts of survival at sea during World War II, there were just 38 shark sightings, only 12 of which resulted in injuries or death.

As reassuring as it was, *Shark Sense* failed to address the most urgent questions on the minds of men afloat in the bedlam of a disaster at sea. Is it true that a shark can smell a drop of human blood in an ocean of seawater? Does noise arouse a shark's curiosity, or scare it away? What about movement? Some accounts—including that of the swimming Ecuadorian—indicated that thrashing scared a shark away; others suggested it sparked their interest. No one really knew.

In 1958, the head of the Biology Branch of the Office of Naval Research, Sidney R. Galler, set out in pursuit of answers. He funded a shark research panel (the Shark Research Panel) and helped establish the Shark Attack File, a database of global incidents that continues today as the International Shark Attack File. David Baldridge's statistical analysis of nine years of Shark Attack File data gave the world—I'm quoting a 2013 National Marine Fisheries Service paper here—"most of what we know today about shark attacks." Much of the rest comes from studies the Office of Naval Research funded in the 1950s on shark predation, olfaction, and feeding behavior. "If you had a good idea for research on sharks," Baldridge told the author of a historical account of shark research, published in *Marine Fisheries Review*, "you went to Sid."

ALBERT L. Tester went to Sid. He had a good idea, he had three species of shark in the ocean outside his door, and he had a pair of fifty-foot-long seawater tanks for experimenting. Tester worked at the Eniwetok Marine Biological Laboratory in the Marshall Islands. (Eniwetok was one of the atolls, along with Bikini,* upon which the US had tested nuclear bombs; the lab provided data on the effects of radioactive fallout on sea life—and, if anyone tracked the obituary

---

* The creator of the two-piece swimsuit, Louis Réard, named it "bikini" because of the explosive reaction he hoped it would generate. The false prefix "bi" has duped many over the years—including the inventors of the monokini, the tankini, the trikini—into wrongly assuming that *bikini* means "two pieces" in Marshallese. In fact, it means "coconut place"—making the term deliciously if inadvertently appropriate.

pages over the ensuing decades, Eniwetok staff.) Tester set out to determine what, specifically, draws a shark to its prey. Do sharks hunt mainly by sight or smell? If it's smell, which smells? Whose smells? If repelling sharks wasn't a reasonable option, a sailor or aviator's best bet was not attracting them in the first place.

Let's start with the good news. Human urine does not attract sharks. When presented with anywhere from a half teaspoon to a third of a cup, blacktip sharks in Tester's tanks took no interest. Neither excited nor repelled, the sharks simply noted the substance, as evinced by a quick turn, or "swirl," which is, I guess, how one acknowledges pee in the pool when one has no eyebrows to raise or shoulders to shrug.

Human perspiration is likewise uninteresting to the shark. It was sufficiently hot and humid in the shark house that Tester and his grad students were able to collect what they needed by sponging each other's bodies and wringing the sponge into a bucket of seawater that was then quietly siphoned into the shark tank. In general, the sharks, and who can blame them, were mildly put off. The perspiration of Albert L. Tester was particularly repulsive to them. At concentrations as low as one part per million, Tester's sweat caused a captive blacktip shark to shake its head and make "a rapid exit from the area."

All-over body sweat—the cooling waters of the eccrine glands—is different from flop sweat. Had Tester done what my friends at the Monell Chemical Senses Center did to me—gathered the pungent armpit exudations of a human under stress—his results might have been different. The sharks might have detected the scent of distress, of easy pickings, and gone into attack mode.

That is precisely what happens when a shark's preferred prey falls under stress. The shark senses a no-hassle meal and closes in to attack. Tester harassed a bucket of groupers by "threatening them with a moving stick" (elsewhere referenced as "poking"). Pumping water from the bucket—scientific nomenclature: "distressed grouper water"—into the shark tank provoked a "violent hunting response." Since the prey were outside the tank, we know it wasn't the sight or sound of grouper pandemonium that set off the sharks' predatory moves. It had to be some chemical exuded through the groupers' skin or gills. And not just any grouper scent would do the trick. When "quiescent grouper water" was introduced into the tank, the sharks paid little heed.

Fish blood and fish guts—two blaring sensory trumpeters of piscine distress—also trigger vigorous hunting moves. So powerful is the chemical signal, Baldridge found, that sharks could be roused to devour a rat—not normally an item of gustatory interest—if its fur were coated with "mullet blend" (whole mullets blenderized with a little water). In a different study, sharks were inspired to attack a kitchen sponge that had been dipped in a bowl of fish body fluids. "Sharks," wrote Baldridge, "will strike essentially anything that has been treated with fish 'juice.'"

That includes spearfishers. In particular peril are those who swim around with the day's catch hanging from their belts or trailing from lines. At the time Baldridge ran his analysis, the Shark Attack File had logged 225 incidents that mentioned the presence of wounded fish and/or fish blood or guts. "Sharks," marveled Tester, "are able to track down and converge on a distressed fish (such as a live fish suspended from a hook through the jawbone but otherwise uninjured) with uncanny speed and accuracy."

Spearfishing probably serves to explain why 17 percent of the Shark Attack File victims were wearing wetsuits. The original theory put forth was that the sharks mistook people in black wetsuits for seals. Perhaps that happens too, but where spearfishing was involved, it's more likely that the wetsuit's accessories—the spear and the belt of oozing fish—drew the shark.

Dead fish also ring the dinner bell. Tester exposed blacktip and gray sharks to a sushi bar of fish flesh: tuna, eel, grouper, snapper, parrot fish, giant clam, octopus, squid, and lobster. All of them he classed as attractants. Sharks prefer to take no risks. They prefer to go after a meal that's not going to put up a fight. Injured is good. Dead is better.

Which makes you wonder about the alleged shark-repellent qualities of decomposed shark flesh. Tester wondered, too. He secured some "alleged shark repellent" from a fisherman, another sample from a fisheries lab, and a sample his team prepared on their own by leaving hammerhead and tiger shark flesh outside in the tropical heat for a week. No repellent effects were observed. On the contrary, it sometimes functioned as an attractant. "Our results . . . seem to be at variance with those of Springer. . . . No convincing explanation can be made." Tester perhaps unaware of the powerful attractant effect of kickbacks from shark-processing plants.

As with fish, so with humans. Over and over, in the shark attack reports of World War II, corpses took the hit. A floating sailor could dispatch a curious shark by hitting it or churning the water with his legs. (Baldridge observed that even a kick to a shark's nose from the rear leg *of a swimming rat* was enough to cause a "startled response and rapid departure from the vicinity.") "The sharks were going after dead men," said a survivor quoted in a popular book about the 1945 sinking of the USS *Indianapolis*, an event that often comes up in discussions of military shark attacks. "Honestly, in the entire 110 hours I was in the water," recalls Navy Captain

Lewis L. Haynes, in an oral history conducted by the US Navy Bureau of Medicine and Surgery, "I did not see a man attacked by a shark. . . ." They seemed to have been, he said, "satisfied with the dead." Haynes says fifty-six mutilated bodies were recovered, but there's nothing to suggest that any more than a few of them were bitten into while alive.

Why, then, do sharks hang around life rafts? For what's underneath. Schools of fish loiter there, either for the shade or to feed on smaller marine life that gathers to take the shade on the raft's underside. Recalled one World War II sailor: "Larger fish came to feed on those minnows, then larger ones to get them; finally the boys with the peculiar dorsal fins arrived to see what the fuss was about." Here's one more, just because I like it: "The shark submerged and swam directly under the raft. . . . We all sat very quiet, . . . and the radar man abandoned the idea of defecating over the side for fear of capsizing. The shark repeated this behavior several times but at no time seemed concerned with us."

And so it continues to be. I know of only one recorded instance in recent history of a shark's biting Navy personnel. In 2009, a bull shark took off the hand and foot—in one bite—of an Australian clearance diver during a counter-terrorism exercise in Sydney Harbor. I asked Naval Special Warfare Command communications specialist Joe Kane about sharks attacking Navy SEALs. "You're coming at this the wrong way," he said. "The question is not, Do Navy SEALs need shark repellent? The question is, Do sharks need Navy SEAL repellent?"

The modern US Navy has no formal shark-attack curriculum. One diver recalls being told to descend slowly and take cover on the bottom should he sense a threat. A 1964 Air Force training film called *Shark Defense* advises downed aviators to blow a stream of bubbles or yell into the water. I asked veteran shark videographer

Robert Cantrell what he thought of this advice. Cantrell has swum among sharks, cageless, for three decades. This is a man who will apply the adjective "nippy" to a group of excited blue sharks. His answer, an answer Baldridge and Tester often came up with, is that it depends on the kind of shark. Screaming into the water may briefly deflect a bull shark, Cantrell notes, but not a tiger shark. Bubbles scare blue sharks, but other species ignore them.

The last Air Force suggestion was a puzzler: "Tearing up paper into small pieces and scattering them all around." I suppose it was meant as a means of distracting the shark—or maybe just the sailor, now absorbed in the challenge of locating sheets of paper while afloat at sea. On one of Cantrell's expeditions, he threw some stale bagels overboard. Tiger sharks swam over immediately; bull sharks ignored them. Cantrell's main advice to the diver who encounters a shark? "Enjoy the experience."

Let us turn now to the question on many a sailor's mind: Is it true that human blood draws sharks? The results of Baldridge's and Tester's experiments are inconsistent. Sometimes the sharks behaved as though attracted to the blood; other times they avoided the test area. Tester wondered whether the freshness of the blood was a factor. In his own experiments, blacktip sharks and greys were strongly attracted to blood less than one or two days old—at concentrations as weak as .01 parts per million of seawater. But Baldridge's analysis of the Shark Attack File data belie this finding. In only 19 of 1,115 cases was the victim bleeding at the time of the attack. "It is difficult," he concluded, "to accept the concept that human blood is highly attractive and exciting to sharks in general when so many shark attack victims have been struck a single blow and then left without further assault even though they were then bleeding profusely from massive wounds."

In Baldridge's own tests, he presented four species of shark with the novel menu option of a swimming, bleeding lab rat. As fellow mammals, rats should possess blood that's about as enticing (or unenticing) to a shark as our own. As he expected, the sharks showed no interest.

The bottom line is that the preponderance of shark attacks, like most animal attacks, are prey-specific. If you don't look or smell like dinner, you are unlikely to be so treated. Predators are attuned to the scents of creatures they most want to eat. Sharks don't relish human meat. Even though a shark can detect human blood, it has—unless starving—no motive for tracking it to its source.

That fact *should* be reassuring to women who enjoy swimming in the ocean but worry about doing so during their periods. But menstrual blood is different, in a uniquely shark-worrisome way. If you'll permit it, a brief shore leave; the US Navy of the 1960s was not interested in menstruating women. The National Park Service, however, was. In 1967, two women, at least one of them menstruating, were killed by grizzly bears in Glacier National Park. Conjecture arose that it had been the blood that inspired the attack. Wildlife biologists didn't buy it, and one of them, Bruce Cushing (delightfully mis-cited in subsequent bear attack/menstruation research as Bruce *Gushing*), set out to collect some data. Cushing opted to study polar bears, because they feed almost exclusively on seals, yielding a clean baseline with which to compare the animals' zeal for menstruating women.

If you put seal blubber in a fan box and aim the aroma at the cage of a wild polar bear, that bear will exhibit what Cushing called "maximal behavioral response." It will lift its head and sniff the air. It will begin salivating heavily. It will get up and pace. It will chuff. It will *groan*. Only one other item that Cushing placed in the fan box could make a polar bear groan: a used tampon. Chicken didn't do the trick, nor horse manure, musk, or an unused tampon. Coming in a close second: menstruating women. The women weren't in the fan box, but in a chair facing the polar bear cage, where they "sat passively," perhaps marveling at the strangeness of life on Earth. Cushing also tested ordinary blood, drawn from people's veins; this elicited no response whatsoever from any of the four participating bears.

In other words, it isn't the blood that makes a tampon attractive to polar bears. It's something uniquely . . . vaginal. Some kind of secretions that, please forgive me, smell like seals. This makes sense, does it not? When a feminine hygiene company hires a lab to test the efficacy of a scented menstrual product, the standardized odor employed for this purpose is known as a "fishy amine."

So alluring is the intensely vaginal/sealy scent of a tampon that a polar bear seems not to notice that it does not also taste like seal. In 42 of 52 instances, a wild polar bear who encountered a used tampon affixed to the top of a stake (scientific nomenclature: "used tampon stake") ate or "vigorously chewed" it. Only seal meat was more consistently pulled from the stake and consumed. Paper towels soaked with regular blood—here again, nailed to a stake like a skull warning foolhardy jungle explorers—were eaten just three times.

What does this tell us about sharks? Should women be worried? Hard to say. How crazy are sharks for seal meat? Do dead groupers smell like used tampons? Unknown. I'd stay in my deck chair, if I were menstruating you.

Cushing concluded his paper by suggesting that since polar bears enjoy used tampons, there was a strong possibility other ursids would, too. But bears, like sharks, vary by species. Forest bears aren't connoisseurs of stinky marine life as polar bears are. Grizzlies like salmon, but they take them fresh. Black bears forage for garbage, so who knows what they've come to develop a taste for over the years.

To settle the matter, here comes the US Forest Service. Had you been off-loading garbage at a certain Minnesota dump on August 11, 1988, you would have been witness to an arresting sight. "We tied . . . [used] tampons to a monofilament line and spin-cast them to foraging bears," wrote Lynn Rogers and two colleagues at the North Central Forest Experiment Station. Despite some fine fly-casting chops on show—the bait being "cast past the bears and dragged back under their noses"—20 out of 22 tampons were ignored. Such was also the fate of used tampons proffered "by hand" to black bears that frequented—though perhaps not anymore—an experimental feeding station. Also ignored: five used tampons tied together and thrown at a group of black bears, as well as all but one of a tasting flight of sodden tampons placed in the middle of a bear trail—four soaked with menstrual blood, one with nonmenstrual blood, and one with rendered beef fat. Ten out of eleven bears "swept their noses closely over the group, ate the tampon containing beef fat, and walked on."

All in all, a resounding testament to the safety of national forests, and the patience of black bears.

FRANK GOLDEN was an authority on the things that happen to a human body immersed for any length of time in cold seawater. Golden—a physician who, by his own description, "swam like a stone"—researched the topic for the Royal Navy Air Medical School during the late 1960s and early 1970s. The text headings in Golden's classic *Essentials of Sea Survival* provide a menu of horrors awaiting service members or anyone else forced to abandon ship or ditch a plane over water: Cold-Shock Response, Breath-Hold Time Reduction, Swim Failure, Drowning, Secondary Drowning,

Saltwater Ulcers, Hydrocution, Trapped Under Ice, Severe Hypothermia, Oil Contamination, Immersion Foot, Turtle Blood,* Sunburn, Wave Splash, Osmotic Diarrhea, Rescue Collapse, Rewarming Collapse. There is no heading for Shark Attack. Sharks don't even make the index.

To a sailor whose sunken craft is a submarine, all of this, the myriad dangers and discomforts of the ocean's surface, are a distant fond dream.

---

* This one is not so bad, provided you know what you're doing. Norwegian shipwreck survivor Kaare Karstaad, whom Harold Coolidge interviewed while working on an ocean survival booklet during World War II, knew what he was doing. He'd catch the turtles at night, when their blood was "cold and refreshing." "Drink it right away," he counseled, "before coagulation takes place." Don't shy away from body cavity fluid! A fifty-pound turtle yields "about 2 cups of 'consommé' which . . . is delicious and not extremely fishy." Sharks, by the way, were "not particularly vicious." (Or delicious.)

12 That Sinking Feeling

When things go wrong under the sea

THERE'S A SOUND THAT water makes, under pressure, when it pushes through a hole too small for its urgency. I know it mainly as a sprinkler sound, a pleasant lawns-in-summer sort of sound. *Phhhhh-hhh....* To a sailor on a submarine, where there are no lawns and no summer, it's not pleasant, this sound. It's the sound of water coming in where it mustn't. A leaking flange, a ruptured pipe. The ocean with its foot in the door. The deeper you are, the harder it pushes. Three hundred feet down, seawater slams through a two-inch hole with enough force to bend a knee the way knees don't bend. At a thousand feet, an eight-inch hole lets an Olympic swimming pool on board every three minutes. If it's not fixed fast, you're in trouble. You're sunk.

I'm looking down into a submarine engine room that's putting out a lot of that sound. Eleven wet necks are bent over leaks—first three, now four. We are 200 feet above sea level, inside a building in Groton, Connecticut, so the risk of drowning is minimal. The room is a mock-up, part of the Naval Submarine School's Damage Control

Trainer, a.k.a. the Wet Trainer, a.k.a. "one of the reasons sailors swear." I'm on the dry side of a large and very clear (it has wipers!) window that looks in on the engine room and the cursing sailors.

With me at the window is the instructor in charge today, Chief Machinist's Mate Alan Hough. Every few minutes he gives directions over his shoulder to a colleague at a console manning the leaks, but his main focus is the students. He's both grading them and giving them feedback. The latter he conveys via signs that he holds up to the window, because no one can hear him through the glass and over the *phhhhhh*. TWO PERSONNEL PER LEAK. WORK BEHIND THE PATCH. NO STRAPPING IN THE WATER STREAM. The signs are rigid red plastic, custom-printed by someone who must have wondered.

Today's subs run on modern technology, but when something goes wrong, the tools sailors turn to may date back to the days of wooden sailing ships. One of the sailors we're watching uses a simple marlin. Beginning an inch below the hole, he winds a length of thin rope tightly around the pipe, choking the leak one wrap-around at a time. The "pine plug" is just a wood cone, an object more commonly seen in building block sets or geometry classrooms. The tip of the cone is hammered into the hole as far as it will go. As the pine absorbs water, which pine does more greedily than most woods, the cone expands, becoming a snugger fit and a more effective plug.

"Horn," Hough says over his shoulder. The man at the console blasts an air horn to make the students look up from what they're doing. Hough grabs a sign (TWO HANDS ON HAMMER) and points at the young man whose hammer and plug the water stream has batted away like a kitten with a yarn ball, or Godzilla with a kitten. This happens nine out of ten times, Hough says; they lose the plug, the hammer, or both. It wastes time when there isn't any to be wasted. And is dangerous. Ninety pounds per square inch (psi) turns a geometry class learning aid into a "pointed missile hazard." The sailor retrieves the cone, which is bobbing on the water a few feet behind him. "One good thing about pine," says Hough. "It does float."

A hammer does not. "That's why we tell 'em: hammer of opportunity." If you lose the hammer, grab what's at hand. This goes equally for plugs. When al-Qaeda blew a 40-by-60 foot hole in the hull of the USS *Cole*, the crew stuffed it with anything they could find. "Mattresses, wood, mooring line, sneakers . . . ," Hough says soberly. "Wrapped it up and shoved it in the hole." It took three days, but they got the flooding under control.

I had met Hough earlier in his office, which he shares with two other men. A jar of Smucker's Goober Grape stood out for the stripey, colorful whimsy it brought to the ill green-beige that someone, at some point, decided to paint the US military. Hough is rangy and pale-complected. He has an appealing overbite that, as he speaks, causes his incisors to touch down on his lower lip like children jumping on a bed. He was raised in a region of the country where people use "them" as an indicator rather than "those." But Hough is nobody's goober. He can take apart a steam turbine faster than most people can put a name to it.

Everything else in today's leak-stopping arsenal is classed as a patch. The term is apt, but misleadingly unintimidating. This isn't like patching a pair of pants. It's like patching a riot hose while the water's still on. You can't come down on the rupture from above. The patch has to be slid over it from the side, like a blanket over a trash-can fire, and then cinched tight.

Hough watches a pair of sailors fail to secure a medium-sized patch called a strongback. The strapping they're using is designed to hold up to water pressure as high as 6,000 psi. "So, for water at

90 psi to be leaking out, that's a very bad job they've done." The red plastic sign Hough would like to hold up does not exist: UNFUCK YOURSELVES.

Hough is tough on his students because the Wet Trainer is a kiddie pool compared to the reality it represents. Here was the situation on the USS *Squalus*, 50 feet down, after a 31-inch air-induction valve failed to close on a test dive in 1939. "The sea had found its way into the maze of pipes that ran the length of the *Squalus*. In the control room, jets of salt water sprayed from a dozen different places." I'm quoting Peter Maas's account of the sinking in *The Terrible Hours*. "The men worked frantically . . . seizing hold of whatever they could to stay upright." And then the lights went out.

And this is from the submarine patrol report of the final patrol of the USS *Tang*, October 24, 1944, the day one of her own torpedoes broached the sea's surface, curved sharply left, and blew a hole in her stern: "The *Tang* sank by the stern much as you would drop a pendulum suspended in a horizontal position." A Lieutenant Lawrence Savadkin described the scene: "With the sudden downward angle of the boat, men and loose gear were bumping and falling by me with the rush of water." The sub school Wet Trainer doesn't tilt, but the one at the Officer Training Command in Rhode Island, nicknamed the USS *Buttercup*, does. (Apparently quite dramatically. "You never save the *Buttercup*," Hough says.) With the understated monotone that comes of hindsight and report-writing, Savadkin concluded, "Confusion was great at this time."

In extreme scenarios like these, the crew skips patches and plugs and heads to the watertight doors. Separating the three or four watertight compartments of a submarine are great, thick round hatches that, in appearance and penetrability, fall someplace between the door of a bank vault and that of a front-loading washing machine.

Everything behind the door may fill, but the flooding stops there. Depending on how much sea has been taken on, an "emergency blow" may be ordered. A blast of pressurized air empties the submarine's ballast tanks like a Heimlich maneuver on a purpling guest. The hope is that this lightening and hollowing of the stricken vessel will counter the weight of the floodwater and float it to the surface.

"If you can't get enough bubble, you're going down." This from Jerry Lamb, technical director at the Naval Submarine Medical Research Laboratory (NSMRL), a few buildings over from the Damage Control Trainer. I've left behind Alan Hough and his sopping sailors to meet with Lamb and one of his counterparts from the UK's Royal Navy, Surgeon Commander John Clarke. Both are well versed in the sequel to damage control: submarine escape and rescue.

Lamb pours me coffee, and Clarke goes off to find milk. He's back a minute later, squinting at the date stamp. "Jan 20. Should be okay."

"What year?" Jerry Lamb is a droll, upbeat soul, his essential good cheer yellowed but slightly by two and a half decades with the Navy. The Navy: smart people, dumb bureaucracy. Meetings, paperwork, conferences. A moment ago I heard Lamb refer to something called the "missile defense luncheon." I pictured doilies under water pitchers and PowerPoints of incoming warheads. Who could eat?

Neither the *Tang* nor the *Squalus* could get enough bubble. The first order of business for a sub on the floor of the sea is to alert potential rescuers. Then, as now, each submarine compartment is equipped with mini launch tubes for flares, smoke signals, location buoys. On World War II–era subs, the location buoy was a sort of floating phone booth in the middle of the ocean. "Submarine Sunk Here," read the sign on the *Squalus* buoy. "Telephone Inside." It was like a *New Yorker* cartoon that didn't quite make sense. There needed to be a third line: "No, *really*." A length of cable connected the buoy to the downed sub. When a rescue vessel arrived, its crew would haul the thing aboard and reach inside for the phone. Peter Maas recounts this moment in his book. The rescue vessel's commanding officer, Warren Wilkin, takes the receiver and opens with a breezy "What's your trouble?" Like he'd pulled up alongside a car on the side of the road with its hood propped open.

The commanding officer of the *Squalus*—here, too, seemingly unflurried in the face of catastrophe—comes back with a chipper "Hello, Wilkie." Whereupon a swell lifts Wilkin's boat and snaps the cable, leaving all further communications to be hammered out in Morse code on the hull of the sub.

Technology has of course advanced since the 1940s. The modern location buoy, SEPIRB (Submarine Emergency Position Indicating Radio Beacon), sends a coded message via satellite with the sub's ID and whereabouts to the closest rescue coordination center. The buoys are still launched through the little tube, though, and ideally that tube hasn't been welded shut, as it was on some cold war–era subs—to keep the buoy from launching accidently and revealing the sub's position to the Soviet subs upon which it was spying. Before a location buoy is launched, someone takes a grease pencil and writes all over it, as much detail as there's room for: damage to the sub (and crew), air quality on board, etc.

What happens next depends on how dire the situation is. Inside every US sub is a fat, white three-ring binder labeled "Disabled Submarine Survival Guide," and in the front of that is a stay-or-go diagram: a decision tree of yes-or-no questions. Is the flooding contained? Are all fires out? If so, if the situation is stable, the answer will likely be Stay. Wait for the rescue vehicle. In water less than 600 feet deep, it may be possible to get out of a sunken sub and make one's way to the surface—Hello, Wilkie!—however, for reasons we'll shortly get to, this is a last resort.

A US sub is stocked with enough oxygen-generating and carbon dioxide–subtracting capability to last at least a week without power: a week of what Clarke calls "bottom survivability." By *bottom* he means the ocean floor, but the British accent, to my ear, anyhow, tilts it toward the naughty meaning. Which kind of fits: bottom as in, "your ass," will it be saved? Seven days is meant to be the outside limit of how long it should take for help to arrive. Fifteen countries and NATO have submarine rescue systems—deep submergence vehicles with decompression capabilities—but they differ in how deep they're able to go. None is designed to function deeper than 2,000 feet; then again, neither are most submarines. (Modern US submarine "crush depths" are classified* information, but educated speculation puts them in the neighborhood of a half-mile down.)

Clarke adds that there may be well more than seven days of supplies. "Because you're probably dealing with a proportion of the crew." It took me a moment to realize what he was saying. He was saying that the oxygen will probably hold out longer than a week, because some of the crew won't be using any. Aboard the *Squalus*, twenty-six men drowned in the first few minutes of the disaster, entombed in the flooding compartment when the watertight doors closed.

The least of anyone's worries is starvation. Subs leave port stocked with full provisions, much of it in cans—so many cans, in fact, that they may overflow the storeroom on the smaller class of

---

* As someone for whom the phrase "top secret" has applied mainly to decoder rings and campy spy movies, I had to remind myself that these are actual security classifications. I found it hard to take seriously the sign on the chain stretched across the navigation room door saying, TOP SECRET—KEEP OUT. They may as well have added NO GIRLS ALLOWED. I saw a printer in the crew lounge labeled SECRET PRINTER. *Secret printer!*

subs, with the result that entire passageways, in the early weeks of an underway, are cobblestoned by cans. Water may be a concern, if the desalination unit isn't functioning. The *Disabled Submarine Survival Guide* includes unflinching water conservation strategies. "Minimize water closet ops following bowel movements to one minimal flushing cycle . . . every three uses." To control odors, the *Guide* recommends covering the mess with the powder used by the galley crew to mix the "bug juice." The high acidity of the drink is pointed out, leading one to assume that that's why it's used for this, though it's also possible it's an editorial comment on bug juice.*

And then you wait. The men of the *Squalus* huddled disconsolately on the torpedo room floor, eating canned pineapple. It is notable that neither crew, *Squalus* nor *Tang*, exhibited panic. Aboard the *Tang*, the commanding officer wrote in his report, "No one was hysterical or disorderly at any time. . . . Toward the last, conversation seemed to be mostly about their families and loved ones." One of the last messages tapped out by the crew on the hull of the sub *S-4*, accidently rammed and sunk in 1927, was "Please hurry." The laborious and time-consuming inclusion of "please" breaks my heart. It's so Navy: courteous and respectful to the end.

The crew of the *Squalus* were as lucky as they were unlucky; the Navy's first submarine rescue chamber had just been completed and tested. The *Squalus* was its maiden rescue. Thirty-three survivors rode it to the surface. The chamber was a modified diving bell. As with an inverted drinking glass lowered into water, the pressure of the air trapped inside keeps the water outside. A diver accompanied the bell to position its opening over the lip of one of the submarine's

---

* Almost as good as this one, by Andrew Karam, in the cold war–era submarine memoir *Rig Ship for Ultra Quiet*: "Bug juice didn't come in flavors, just colors."

hatches and bolt it in place. The sub's hatch could now be opened, and small groups of crew helped into the chamber.

Prior to this, sinking was likely a death sentence. Even a few inches of water will bear down on a submarine hatch—or a car door, for that matter—with sufficient pressure that it can't be pushed open (unless one equalizes the pressure by letting water in). On the smaller subs of the 1920s, the air would last about three days. It was one of these "iron coffins," the S-51, that inspired Lieutenant Commander Charles "Swede" Momsen to come up with a way to get people out. Momsen's sub had been the one that first arrived on the scene. All the crew could do was stare at the oil slick on the surface of the water, "utterly helpless," as Momsen wrote to a friend. When the sub was salvaged, bodies of the crew were said to have been found with their fingers torn and bloodied from trying to pry open a hatch against fifteen tons of ocean.

Given that most US ballistic missile submarines today spend the bulk of their time in oceans that bottom out deeper than their crush depth, the term "iron coffin" has regained some accuracy. Crush depth is the point at which the hull succumbs to the extreme water pressure and the sub implodes. John Clarke likens it to putting a submarine inside a giant bomb. The sub shatters inward. And the crew? "If you can imagine," Clarke says, "all the metal parts are imploding together and anything in the way would be crushed and shredded and pounded into bits." No one saves your bottom now. On April 10, 1963, the USS *Thresher* imploded, killing all 129 men aboard. "She's scattered all over the seafloor," Jerry Lamb says.

In light of the deep-sea haunts of modern subs, why even bother with rescue and escape systems? Do they simply exist to, as one submariner expressed it, "give moms and dads a warm feeling"? No, no more so than airplane emergency exit slides. Because, as with airplanes, most collisions take place on arriving or departing: in port or airport, where the traffic is busiest but the plummet most survivable.

The *Tang* went down in water just 180 feet deep, but rescue was complicated by the circumstances of battle. She sank in the midst of the convoy of Japanese ships she'd spent the night torpedoing and sinking. In the end, bad air forced the crew's hand. Smoke had built up from the burning of classified paperwork, and saltwater had reached the batteries, creating deadly chlorine gas. Disaster luncheon. You didn't need a decision tree to know that Stay had turned into Go.

Swede Momsen invented something for this scenario, too. During World War II, subs were equipped with escape trunks and Momsen lungs. (The "lung" was a wearable air supply that, upon reaching the surface, handily converted to a flotation device). Like an airlock on a spaceship, the escape trunk allows for the equalization of pressure inside and outside. On a sub, this allows the hatch to be popped open and the lung-clad sailor set loose in the brine. The *Tang* was the first bottomed sub from which sailors escaped without the aid of a rescue bell. There were nine, four of whom subsequently drowned or disappeared. (In the surreal etiquette of war, the five survivors were plucked from the near-freezing water by their enemies—as the *Tang*'s commanding officer described them, "the burned and mutilated survivors of our own handiwork"—who then beat them and sent them to starve in a prisoner of war camp.)

What happened to the rest of the men gathered in the *Tang*'s torpedo room in their Momsen lungs? Why didn't they escape? They weren't sure how to do it. "A majority of the men," reads the patrol report, "had never been properly trained in the use of the Momsen lungs or operation of the escape tank. They, therefore, didn't have any self-confidence in their ability to escape, causing a general feeling of defeat among them. . . . After the first two attempts there were very few men left who cared to try an escape although they knew what was going to happen to them below." They are all there in the Summary of Escapes. Torpedoman's Mate Fluker: "Would not try after this, his second attempt." Unnamed Ensign: "Removed in stupor from trunk; preferred not to try again." Unnamed Machinist's Mate: "Would not try after this, his first attempt."

A little practice might have made the difference. "Although everyone had read how to escape," says the report, "not one had actually went through the motions." In 1930, at the urging of Swede Momsen, an escape training tank was commissioned for the submarine base in Groton. With the hope that every submariner would have a chance to went through the motions.

A T 40 feet deep and 84,000 gallons, the Naval Submarine School's Pressurized Submarine Escape Trainer holds easily as much water as a hotel swimming pool. In diameter, though, it's closer to a Jacuzzi. It's the sort of thing you might drop into by accident, like a manhole, because you didn't notice it was there. Despite the aquamarine water and the echoey tile walls, *pool* isn't the right word. This is a column of pretend ocean that exists for a single, highly nonrecreational purpose: to practice bailing out of a stricken sub.

Twenty-six sub school students stand around the water's perimeter in identical (Navy) blue swim trunks. They are young enough that the pimples on their backs still outnumber the tattoos. In ten years it will be different. Navy boys accrue ink like sun damage. A little more every year, in every port. The first training exercise will begin in an escape trunk that feeds into the water fifteen feet down. No breathing apparatus will be worn, just a life jacket. The instructor calls it "a buoyant exhaling ascent," a term I will tuck away for later use should I ever be called upon to write opera reviews.

Exhaling is the word to be underscored. Faced with an ascent from deep underwater, novice swimmers are inclined to hold their breath—not just to stay alive, but to help buoy them to the surface. They may not realize that that initial lungful of air they took in will expand as they rise and the water pressure decreases. If that breath expands enough, it will burst the lung's alveoli—the tiny sacs where an exchange of gases in the air and the blood takes place. Should this happen, air bubbles can get into the bloodstream. Air embolism. Not good. Critical care luncheon. The bubble can act like a clot, blocking blood flow and starving organs of oxygen. If the organ in question is the brain or heart, the tissue damage may be fatal. There is speculation in the *Tang* patrol report that this had been the fate of four men who made it out of the escape trunk but then disappeared: that they'd lost the mouthpieces of their Momsen lungs and hadn't realized the consequences of holding their breath.

"It's the Golden Rule of sub school," the instructor, Eric Nabors, is saying. "Don't hold your breath." Nabors carries the evocative title Diving Officer, and seems built in keeping. His hair is buzzed to a half millimeter, his wedding band tattooed. Nothing disrupts the hydrodynamic flow of Eric Nabors in a wetsuit.

To modulate their exhalation—not too fast, not too slow—the young men are instructed to pretend they're blowing out birthday candles. Yelling also works. To further discourage breath-holding, Nabors and his fellow instructors used to inflate a wine bag down at the bottom of the water and let it go. As it surfaced, the bag would burst.

While Nabors and I have been chatting, I've referred to the wine bag as a bota bag. Nabors finally stops me. "*What* are you saying?"

Did I have the wrong term? That goatskin pouch that herders used to sling over one shoulder? In Spain? The kind where you open your mouth and squirt in the wine?

Nabors blinks at me. "I'm talking about the bag from wine-in-a-box."

My escort for the day has been chatting with Nabors, and I notice she calls him "Jim." This would explain the Jim Nabors album (*Kiss Me Goodbye*) mounted on his office wall, but not the ID badge, which says "Eric Nabors."

"I fought that battle for a long time," he says when I bring it up. When your last name is Nabors, there will be people who call you Jim, no matter what you do to discourage them. "Eventually I gave up."

The bursting bag has been replaced by a video of itself, because the real thing was too intimidating, and no one wanted to get in the Escape Trainer afterward. Few of the students will cop to it, but there's some anxiety in the house today. Some of these boys can barely swim. The Navy entrance requirement is minimal. You are dropped in a pool fifty feet from the edge, and you get to that edge however you can. You don't have to like water to join the Navy. "I don't even like baths," said one submariner I met.

Nabors explains to the students the sequence of events. A pair of divers are with each student up to the time he begins his ascent, to be sure he's exhaling at the right rate, that he's been able to clear his ears, that he's not feeling panicked. Then they let him go. It's over in a few seconds. "You're going to pop out of the water, and a diver is going to say, 'Are you okay?'" Nabors says. "And you're going to shout your name, your rank, and 'I'm okay!'" (So the guy standing by with the clipboard can put a check next to the name.) "Got it?"

"Yes, *sir*!"

A few minutes later, the first student pops out of the water, buoyant with air and relief. A diver is there to receive him and steer him to the edge. If you wandered onto the scene without knowing where you were, you might think, *baptism?*

"Are you okay?" shouts the diver.

"Yeah." Nabors and the clipboard guy exchange a look. Kids today.

One student backs out of the ascent. You can tell who he is by the red bathrobe; everyone else's is tan or blue. This isn't done to shame him; no one but the staff knows the significance of being "red-robed." It's a way to alert them to keep a watchful eye out, in case a medical issue develops. In this case, the boy was just scared. He confesses a fear of drowning. I glance at his bare feet for the traditional Navy "anti-drowning tattoos": permanent inkings of a pig and a chicken, one on each foot. Because when the old frigates sank, pigs and chickens from the ship's hold could be seen floating on the water's surface.

The boy's fellow students were sympathetic, and this he expected: "One team, one fight." I've heard the word *brotherhood* applied to submariners. At 7 percent of the Navy, it's a tight-knit community. Especially boat by boat. Where an aircraft carrier crew may number 6,000, US submarines have room for fewer than 200. There's an intimacy born of not only the diminished personal space that the smaller classes of subs impose but the months-long isolation and, until recently, the absence of women. "There's a lot of hugging and stroking heads," a former NSMRL psychologist told me. "I was taken aback by how physically affectionate they are."

Inevitably, this has fueled rumors. Andrew Karam, author of *Rig Ship for Ultra Quiet*, told me about sitting in a bar with his shipmates when a "skimmer"—a surface sailor—walked in. "When he realized we were all submariners, he said, 'I know about you guys. Hundred forty men go down, and seventy couples come back up.'"

"That's not true," Karam deadpanned. "We have some threesomes."

The US Submarine Force began integrating female officers in 2010, with enlisted ranks following in 2016. So far, so good. Jerry Lamb says a recent ban on cigarettes provoked more clatter. And then this happened: The day before my visit, *Navy Times* broke the story that female officers on the USS *Wyoming* had been filmed in the shower.

I ask Nabors whether he has to tell his students not to urinate in the Escape Trainer.

"It's not even a topic of discussion. It happens."

I forgot he's a diver. I'm told divers pee in their wetsuits. Me, I've never. "I can't even pee in the ocean."

The guy with the clipboard glances at Nabors. *Wow*, the glance says. *Live a little.*

THE STUDENTS troop single file to the stairwell, ducklings in a row. They are going down to the bottom of the Escape Trainer for the big ascent, the 37-footer. This time they'll wear a SEIE (Submarine Escape and Immersion Equipment) suit, a partially inflatable head-and-face-encompassing zip-up that attracts unwanted comparisons to a body bag. Like its more minimal predecessors the Momsen lung and the Steinke hood, the SEIE suit incorporates an air supply and openings to vent excess air as it expands on the ascent. This allows the escapee to breathe normally and not have to worry about bursting lung parts. Students practiced the "exhaling ascent" earlier so they'd know what to do if there was a problem with the escape suits. Something like this, for example: "The rubber was cracked and tacky and most of them were stuck together." This is Andrew

Karam in an email to me, describing the Steinke hoods he was asked to inventory for an underway in the late 1990s.

"On the bright side," Karam went on, "we spent almost all our time in water more than a thousand feet deep, so opportunities to use them were few and far between." The maximum depth at which a Steinke hood has been successfully tested is 450 feet. The greatest depth at which a SEIE suit can be counted on to save you is 600 feet. That is really all you need ask for, because if you're escaping into water deeper than 600 feet, you're likely to be killed by decompression sickness no matter what outfit you have on.

To understand decompression sickness (the bends), it's useful to think about one of those countertop carbonation units. Bubbly water is tap water with the bends. When you force pressurized gas into a container with liquid in it—be that container a SodaStream bottle or a scuba diver—some of that gas may go into the liquid. (To get all jargony, the gas goes "into solution" for the greater cause of equilibrium.) Now say the pressure in the container lets up suddenly—because the bottle has been opened or the diver has swum up toward the surface. Those gas molecules that had been forced by the air pressure into the liquid will now come back out of solution. (Here again: seeking equilibrium.) As they do this, the gas molecules link together in the form of bubbles. Never mind why. They just do. Now you have a glass of refreshing fizzy water, or a looming case of the bends. The bends is bubbles migrating through the body and causing problems: acting like a clot and disrupting the flow of blood to vital organs, or pushing apart tissue and causing pain, or both, and more.

Divers can avoid the bends by ascending slowly. This gives the body a chance to simply exhale the gas as it comes out of the blood and into the lungs. (Nitrogen is the main culprit; air contains a lot of it, and it likes to dissolve and hide in fat.) The more time a diver has spent breathing pressurized air, and/or the more highly pressurized the air, the more nitrogen she'll need to dump and the slower, therefore, she'd need to ascend.

Decompression may or may not pose a danger to escaping submariners. If they're lucky, the air inside the stricken sub has remained as it was when they left port: pressurized to sea level. In that case, submariners can usually escape with little danger of the bends. But if the vessel floods, the water that's come on board will compress the air like a trash compactor. Now the sailors are like scuba divers: They're breathing pressurized air, and some of the gases in that air will be pushed into their blood and tissues. Depending on how long they breathe this air and how compressed it is, they may, like a diver, need to decompress in order to ascend to the surface safely. Breathing the pressurized air in the escape trunk for the minute or so that one is inside it isn't enough time to create a problem unless one is down very deep. At, say, 800 feet, the air in the escape trunk would have to be so highly pressurized (to equalize with the outside pressure and allow the hatch to open) that breathing it for even a minute would force enough nitrogen into the body to put one at risk for the bends.

At the far, nightmare extreme of the bends is something called explosive decompression. On November 5, 1983, four divers were relaxing in a decompression chamber on the deck of an oil rig in the North Sea. For reasons that remain unclear, one of the dive tenders opened the hatch, reducing the pressure in the chamber from what it would be at 305 feet underwater to what it is at sea level—in a fraction of a second. Nitrogen bubbled out of solution instantly, in the men's brains, their blood, their fat and muscles. The pathologists wrote in the case report that the men's fat looked like "sizzling butter on a frying pan." They surmised that the blood had begun to bubble instantly, "leading to an instantaneous and complete stop of the circulation."

Diver four had been at the hatch when it blew. Good-bye, Wilkie. He was the champagne at the top of the bottle. The pathologists speculated that in addition to the breakage caused by being shot through the partially open hatch, he "also must have exploded." He arrived at the autopsy suite in four plastic bags. Some of his organs were missing, having been "blown straight into the sea." Like the gases in the abdominal cavity, air in the brain pan had also expanded in an explosive manner. "The scalp, with long blond hair, was present but the top of the skull and the brain were missing."

I've been looking through a porthole at the bottom of the Escape Trainer. Watching this scene, the rag doll movements of the divers and the silvery jellyfish air bubble that floats from the hatch in a languid blurp, it is easy to forget the murderousness of deep water. Submariners can't afford to forget. Mistakes can so swiftly give rise to disaster, and then where are you? Too deep for help or escape.

The risks are amplified when someone on your crew has been up for sixty hours. It has been an enduring unease of the military that the last people you'd want operating subs or fighter jets or automatic weapons are often the ones doing it: the persistently, catastrophically sleep-deprived.

13 Up and Under

A submarine tries to sleep

THE BEDS IN THE missile compartment are a recent addition. When the USS *Tennessee* got a technology upgrade, some years after the sub was built, extra crew were needed to serve the servers. This posed a problem, until it occurred to someone that there's room for a bunk pan in the space between two nuclear missiles. The Trident II launch tubes—of which there are twenty-four on board—stand 45 feet tall, spanning all four decks of the submarine. The multifloor missile compartment is the least hectic place on board. It's like the stacks in an old college library—a still, private place to put your head down and catch some sleep.

Though not just now. "All hands awake!" The voice on the intercom is accompanied by an alarm, loud and insistent. *Bong, bong, bong, bong.* An annoying child with a stick and pot.

"Simulate sending all missiles." This is a lot of missile. Each Trident carries multiple warheads, each programmable to its own destination, with sufficient precision to, as I've twice heard it put, "hit a pitcher's mound." The ballistic missile submarines of the US fleet, fourteen in all, are a roving underwater nuclear arsenal. Along with missiles in underground silos and others on bomber aircraft, they make up the "nuclear triad" of US strategic deterrence. You would be crazy to nuke us, is the message here; we have more bombs than you have, and you can't take ours out first because you'll never find the ones on the subs. Ballistic missile submarines have whole oceans to hide in, and a nuclear reactor aboard to generate power and water, so they never need to surface for fueling. They can stay deep until the food runs out.

The *Tennessee*'s second-in-command, Executive Officer Nathan Murray, invited me to join him in the missile compartment for the drill. (I sailed out to the sub with a group of prospective command-ing officers going aboard for a practical evaluation.) We pass a row of sleeping spaces along one wall, some closed off with black vinyl curtains, giving them the look of bathroom stalls at certain punk clubs in the 1980s. Murray points out the bed of a young man who shares his space with the wall coupling for the fire hose. He was woken up last night for a fire drill, and now this.

The Submarine Force has formally acknowledged that it has a sleep problem. Quoting *Force Operational Notes Newsletter (Special Crew Rest Edition)*, "An individual's sleep at sea is not protected, allowing administrative training, maintenance, and 'urgent' matters to routinely shorten or interrupt a person's sleep. . . ." The crew of the *Tennessee* endure fire drills, flooding drills, hydraulic rupture drills, air rupture drills, man overboard drills, security violation drills, torpedo launch drills. They practice launching the missiles more regularly than some people floss. On one hand, you want the crew to be well trained. You don't want to hit the wrong pitcher. On the other, you don't want training and drills going on so often that the people tending the bombs and reactors are chronically sleep-deprived.

In 1949, submarine schedules allowed ten hours a day for sleep. On top of their "long sleep," half the crew took at least one nap. Starting in 1954, subs went from diesel to nuclear-powered engines. The result being that there's a lot more to watch than a temperature gauge and an oil level. On the USS *Tennessee*, four hours' sleep has been about the average.

Before coming aboard, I spoke by phone with sleep researcher Colonel Greg Belenky (Ret.), the founder of the Sleep and Performance Research Center at Washington State University, Spokane. Belenky knows what happens when people go from sleeping eight hours a night to sleeping four or five. Their cognitive mojo declines over the course of a few days, whereupon it plateaus, settling in to a new, compromised state. The less sleep they're getting, the longer their mental abilities deteriorate before they plateau. Which mental abilities? Most. Sleep deprivation shrinks memory and dims the network that sustains thought, decision making, and the integration of reason and emotion, Belenky said. "You know when you have a problem you're working on and you give up? Then you get a good night's sleep, wake up, and suddenly there's your solution? That's what sleep does. It returns the brain to its normal specs."

On submarines, the junior crew have it worst. On top of work and watch duties, they are preparing for "qualification," a sort of submarine version of passing the bar: sixty-plus verbal quizzes on submarine components and systems plus practical tests on various elements of your particular sub—anything from taking the helm to using a fathometer to blowing a sanitary tank. "I'll get three hours of sleep one night, and the next night none," said a long-faced seaman studying dive hydraulics in the vaporizer haze of the *Tennessee*'s enlisted crew lounge. (Between the vaping, the zombie-apocalypse video-gaming, and some aggressive tabletop football flicking, a *terrible* place to study. Or maybe just to be middle-aged.)

The seaman will tell you he's fine, but Belenky knows he's not. When people drop below four hours a night, they don't plateau. Their abilities continue to erode until they end up at the point where sleep researchers have had to come up with special terms, like "catastrophic decompensation." "Put simply"—and here *Force Operational Notes* shifts into typographic overdrive, simultaneously boldfacing, underlining, and italicizing—"failure to get adequate continuous sleep every day results in overly fatigued personnel who, in a matter of days, function at a deficit similar to being intoxicated."

Like drunks, the chronically sleep-deprived are doubly dangerous in that they're poor judges of their own impairment. Jeff Dyche, a sometime research psychologist at NSMRL, now with James Madison University, told me about a study that showed that people who'd slept six hours a night for two weeks were as cognitively diminished as people who'd been up for forty-eight hours straight. Unlike the up-all-nighters, routine six-hours-a-nighters see no need for caution. They've felt mildly exhausted for so long it's become their normal, Dyche says. "They're like, 'Ah, I'm used to it.'" I've been hearing a lot of this over the past two days. "I get four and a half hours and I'm generally okay for a twenty-four-hour period," said a sailor pushing trash into an institutional-grade compactor that would work with equal efficiency on flesh and phalanges.

Murray and the sub's commanding officer, Chris Bohner, volunteered to try out a new watch schedule aimed at keeping crew better rested, both for their health—insufficient sleep having lately been linked to obesity, high blood pressure, diabetes, heart disease—and for everyone's safety. It is not a simple undertaking. "I spend a very significant amount of time," Murray says, "figuring out people's rest." Murray is a popular leader—in both manner and mien, a solid individual. You never see him slouch or lean or jut one hip. He stands steady and square on both feet, like a bag of mortar set down. His hands park on his belt, with an occasional sweep over his head, which he keeps closely shaved. The latitude of Murray's hairline, like that of the submarine itself, will remain a secret to me.

The problem is that things come up. People fall behind and schedules fall apart. The problem this week is me. Everyone's work was interrupted because the crew had to spend four or five hours looking for a spot where the seas were calm enough to drop a gangplank between the sub and the vessel we sailed out on.

Part of the Navy's challenge in dealing with undersleeping has been that somewhere along the line, it became a point of pride. At NSMRL I met a longtime submarine commanding officer named Ray Woolrich. "Marines sitting around in a bar," said Ray, "will tell you how many push-ups they can do. Aviators will tell you how many g's they can take. Submariners will tell you how many hours they stayed up." Better to be exhausted than to gain a reputation as a "rack hound."*

For decades, military sleep research proceeded in lockstep, focusing less on getting sleep than on getting by without it. Study after study tested this or that stimulant on fliers, soldiers, sailors. Only recently has protecting sleep become a Defense Department priority. Current Army policy requires unit leaders to develop and implement a sleep management plan in theater. (Though in one small survey of soldiers returning from Iraq or Afghanistan, 80 percent had never

---

* In military slang, there's a friendly epithet for everyone. I, for example, am a "media puke."

been briefed on such a thing.) A turning point, according to Belenky, was the lengthening of the Army's field training exercises (FTXs), the massive simulated confrontations that serve as a sort of practical final exam for soldiers. "At some point the doctrine folks had concluded that any war worth going to would probably last a week or two, so they increased the duration of the FTX from three days to two weeks," Belenky said. Up to that point, there had been a tradition of staying up for all of it, in order to "look motivated and get a good evaluation." Belenky recalls getting a call from a commander shortly after the change went through. "He said, 'I need your advice on pharmacology. I need my boys to be able to stay up longer.'" Belenky figured the man was talking about a couple extra days. "I said, 'How long do you want them to stay up?' He said, 'Two weeks.' People actually tried to gut it out." It was a vivid and no doubt fairly entertaining demonstration of the importance of sleep to military competence.

History provides equally vivid demonstrations. Medical historian Philip Mackowiak compared eyewitness and officers' accounts of Stonewall Jackson's performance during a series of Civil War battles with the general's opportunities for sleep, if any, in the days leading up to those battles. In 100 percent of the battles for which Jackson had had no chance to sleep in the three days prior, his leadership was rated "poor." In the Battle of Gaines' Mill, his chief of staff described him as "thoroughly confused from first to last." His brigades were not merely "out of order"; "he did not know where they were." The Battle of Glendale found Jackson "benumbed, incapable . . . of deep thought or strenuous movement . . . uninterested and lethargic." At times during the Battle of Malvern Hill, Jackson "appeared to be almost a bystander." In the midst of the Battle of McDowell, he was discovered napping.

For every twenty-four hours awake, Belenky told me, people

lose 25 percent of their capacity for useful mental work. Jackson was leading the charge (or not) on 25 percent of his waking best. I'm trying not to think about a man named Patterson in one of the *Tennessee*'s machinery rooms. He'd been up for 22 hours trying to fix the electrolytic oxygen generator, a large, pulsing metal-hulled molecule splitter. "Basically it's a hydrogen bomb," he'd said cheerfully.

The longest Belenky has kept subjects awake is 85 hours—three-plus days—which is about the limit, he says. "They're not," he adds, "very useful to anybody." There are people who claim to have stayed awake for 100 and even 200 hours, but because their brain waves weren't continuously monitored, as Belenky's subjects' are, it's impossible to be sure they weren't microsleeping. The very tired can slip into Stage 1 sleep for a few moments, eyes open, carrying out some quasi-coherent version of whatever it is they're up to. As anyone who has slept on an airplane knows, it's possible to maintain muscle tone while sleeping—that is, until you slip into REM sleep, during which muscles relax. (When people fall sleep at odd times in their circadian cycle, they may enter REM early. Blame "early-onset REM" for the slack-jawed head-lolling that happens when you nap sitting up.)

Soldiers, including Stonewall Jackson's, have on occasion reported sleeping during night marches. If you're tired enough, Belenky says, your brain appears to briefly dissociate—one part sleeping, another awake. There are birds and marine mammals that manage this regularly. Dolphins and seals are able to sleep unihemispherically—with one half of their brain. This is because the other half needs to attend to breathing, which in their case requires swimming to the surface for air. When geese and ducks sleep in groups on the ground, the birds on the outer edge will keep one eye open and the corresponding brain hemisphere awake, scanning for predators.

From a military perspective, a soldier who could march or swim or look out for enemies while simultaneously catching up on sleep would be a desirable item. It fits right in with one of the goals of the military's futuristically minded Defense Advanced Research Projects Agency (DARPA): "to enable soldiers to stay awake, alert, and effective for up to seven consecutive days without suffering any deleterious mental or physical effects and without using any of the current generation of stimulants." This is why you'll find the Defense Department on the sponsor lists of some of the basic research on unihemispheric sleep. If science could just figure out how the ducks do it, perhaps troops could be enabled—chemically or surgically, God only knows—to do it, too. Belenky scoffed. "We're not even sure what triggers *whole brain* sleep."

That hasn't stopped military organizations from fantasizing about it. I came across a NATO symposium on Human Performance Optimization that included a roundup of medical technologies that might be repurposed to optimize warfighters. In among the prosthetic limbs "to provide superhuman strength" and the infrared and ultraviolet vision–bestowing eye implants was this: corpus callosotomy to "allow unihemispheric sleep and continuous alertness." Surgeons have on occasion severed this connector between the brain's halves as a way of reducing the number of seizures in patients with severe epilepsy. Does this in fact change how these patients sleep? No, says Selim Benbadis, director of the University of South Florida Comprehensive Epilepsy Program and the author of a paper on the procedure. He added that there are infants with incompletely developed corpora callosa and they sleep normally and with both hemispheres at the same time.

"They think a lot of harebrained things are good ideas," Belenky said of DARPA. Yes, they do. The wish list also included "surgically provided gills."

• • •

"**R**ELEASE OF nuclear weapons has been authorized." It's the intercom man again. Even in a simulation, it's a sickening thing to hear. I look around at the sailors standing near. One untangles an extension cord. His face betrays nothing. A sailor seated at a control console blows his nose. "Is this what it would be like?" I ask Murray. "If it were the real deal? Would people just be calmly carrying out their tasks, blowing their nose . . ." The whole business is straight off my fathometer.

Murray's not playing this game. "If your nose is running, you blow it."

Two sailors hustle past, each holding a corner of what looks at a glance like some kind of Lotto ticket. It's the code for the key box, the box with the keys to launch the missiles. Two people must have a hand on it at all times once it's out of safekeeping, for the same reason some airlines, in the wake of the 2015 Germanwings suicide flight, require a second person in the cockpit.

Were this an actual missile launch, I'd wager that adrenaline would keep the crew alert regardless of how long they'd been up. But the normal day-to-day routines of a ballistic missile sub are a good deal less invigorating. Most watches are just that: hour upon hour of watching. Watching displays, readouts, dials, sonar feed. It's a worrisome mix: sleep deprivation, tedium, and large, potentially destructive items. "The Navy doesn't want us to publish anything saying that these guys monitoring these nuclear reactors are falling asleep on watch," Dyche told me. "But we know that's happening." Even awake, the tired are poorly suited for standing watch. When psychologists give sleep-deprived people a standard battery of cognitive tasks, their score on measures of "psychomotor vigilance"— paying attention and noticing shit—drops dramatically.

I never visited the *Tennessee*'s reactor and its tenders, because I didn't have security clearance for that part of the sub, but I did visit the torpedo room. There are four of them, massive as medieval battering rams. Sweetly (I guess), they are named for the torpedomen's wives. I asked the torpedoman on watch when last a US submarine had had cause to fire a torpedo at another vessel. He thought for a moment. "World War II." He's the Maytag Repairman, ready for action in the extremely unlikely event it's called for. The torpedoman's watch is a checklist of inspections, walk-arounds, paperwork. Always with the paperwork.* Outside of the sonar shack and the Missile Control Center, much of the *Tennessee* remains charmingly analog. I looked around the missile compartment at one point and thought, *tuba parts*. The torpedo launch console has big square plastic buttons—Flood Tube, Open Shuttle, Ready to Fire—that flash red or green, like something Q would have built into James Bond's Aston Martin. The missile compartment has similarly retro-looking panels of buttons. They provided the setup for one of the more quotable things Murray said to me—a line that, were fewer precautions in place, could have joined "Houston, we've had a problem" or "Watch this" in the pantheon of understated taglines for calamity: "I wouldn't lean on that."

On an intuitive level, the prospect of marginally vigilant humans babysitting reactors, torpedoes, and weapons of mass destruction is unsettling. That the scene takes place in a vessel under hundreds of

---

* By weight, a submarine carries more paperwork than it does people—despite the best efforts of Vice Admiral Joseph Metcalf III. Metcalf, who led the invasion of Grenada, waged an equally headstrong campaign for shipboard computerization—"a paperless ship by 1990," he told the *New York Times* in 1987. He calculated that even a smaller surface warship carries 20 tons of technical manuals, logs, forms, and shelving—tonnage that could be used for fuel or ordnance. Metcalf's battle cry ("We do not shoot paper at the enemy") attracted some media attention and probably one or two spitballs, but—if the USS *Tennessee* is any indication—no serious commitment to change.

feet of water, all the more so. Statistically, however, the highest risk doesn't lie in the nuclear reactor compartment or even, for that matter, in deep water. The biggest risk lies with the seemingly straightforward but in fact reliably harrowing task of surfacing a sub.

A BALLISTIC MISSILE submarine will take you to the remotest places you'll ever travel and show you none of it. The sub has no windows or headlights, nothing to make it visible in the surrounding black. Below the depth that sunlight penetrates, a periscope is useless. The crew see by sonar, picking up propeller sounds from ships and plotting their distance and course. To remain undetected, ballistic missile subs use passive sonar only: no pinging. Echolocation—sending out sound and timing its bounce-back—would give away the sub's own location. The *Tennessee* is blinder than a bat.

At 450 feet down, our current depth, there are no other vessels to smack into. (Each sub has an assigned territory, or "box," extinguishing the infinitesimal likelihood that two of them might collide.) The biggest danger outside at the moment is shrimp. When galley crew empty the grind bucket, vast schools of snapping shrimp rush the hull to feed. Their collective tumult can mask engine noise from other vessels.

In the sonar shack this morning, four men sit at monitors, watching snowy green crawls of sonar feed and listening through headsets. A sonarman can identify a ship by propeller noise the way a birder might distinguish one woodpecker species from another by the speed or timbre of the hammering. Someone passes me a headset to hear the click-jabber of some porpoises. After a few days in a submarine, any contact with nature can be a bit heady. "Flipper!" I hate to apply the verb *squeal* to myself, but that's what it is.

"Uh huh," says the sonarman. "Flipper *all night long.*"

Although ballistic missile subs are able to stay deep for months, they typically do not. The *Tennessee* surfaces regularly, like a whale, to exhale emails. We're about to come up in a merchant transit lane, which has everyone a little on edge. In the hour-long lead-up to the moment when the sub breaches the water's surface, someone's been at the periscope, face pressed to the eyepiece, scanning for anything sonar might have missed. Because the view is less than 360 degrees, he circles slowly, around and again, crossing one leg behind the other, a slow dance with a canister vacuum. You want to be very, very sure there's nothing up there.

In 2001, the USS *Greeneville* surfaced directly beneath a 191-foot Japanese fishery training ship. The sub's rudder sliced the hull, causing the trawler to sink and resulting in the death of nine people aboard. (Sleep deprivation wasn't cited as a contributing factor. A group of visitors was: fourteen CEOs and, um, a writer. All but one of the group were up in the control room, crowding the periscope platform, blocking access to critical displays, distracting the sonarmen.)

The captain of the *Greeneville* exhibited what is known in these parts as poor periscope discipline. He scanned for about half as long as procedure called for. Another potential danger for a surfacing sub is "bow null." If the front of a ship points straight into a submarine's sonar array, the sound waves emanating from that ship's propulsion are blocked by its own body and cargo. The *Tennessee*'s safety officer compares it to "yelling through the trunk of a car to your kids in front of the car." A helpful, if disquieting, metaphor.

It's the weekend, which can be a more dangerous time to come up. Container ships that are nearing a port outside standard work-week hours will sometimes loiter, timing their arrival for Monday, when pay scales drop back to normal. A container ship is the size of a strip mall, but if its engines are silent, it's all but invisible to the crew of a ballistic missile sub. Aboard the *Tennessee*, a sailboat is more worrisome than a warship. Now you understand how it came to pass that the USS *San Francisco*, in January 2005, ran into an undersea mountain. They're very quiet, mountains.

Adding to stress levels: Last-ditch evasive maneuvers are out of the question. A surfacing ballistic missile sub is traveling between 6 and 12 miles per hour. "It's like a baby crawling out of the way of a truck," says the safety officer, as though yelling through the trunk of a car that there may be something just a little bit off with him.

Extreme caution is ever the mind-set. If a new sonar contact should appear on the screen during surfacing, an "emergency deep" may be ordered. Because without echolocation, you don't immediately know how far off the other vessel is. "Be safe now and figure it out after," the commanding officer said yesterday, as we dove to avoid a ship that would turn out to be several miles off. A ballistic missile sub is a boat without a destination, its course a series of evasions and nervous retreats. Any time a contact is calculated to be within two miles, the commanding officer is called. And, often, the navigator and the executive officer.

And there goes another night's sleep. "I expect to be woken three or four times per sleep," the navigator told me. Murray wakes up, too, because he has a speaker mounted on the wall of his stateroom, above his pillow, that picks up the conversation in the control room. He's like a new mother with a baby monitor on the nightstand. "All of a sudden, out of a lot of background noise and chatter, you'll catch a certain word or a change in the tone or volume of somebody's voice. It just snaps you out of a sleep."

Unsurprisingly, submariners have a robust tradition of caffein-ation.* The *Tennessee* left port with a thousand pounds of coffee. The world's first nuclear-powered submarine, built in 1954, is now a floating museum in Groton, Connecticut, and if you tour it, you will see metal rings bolted to consoles and bulkheads at the different watch stations: cup holders! Caffeine is safe and effective but not without a downside. Depending on one's sensitivity, it has a half-life of six to eight hours. Even if you have no trouble falling asleep after drinking coffee late in the day, you may wake more easily during the night because your nervous system is still aroused, your brain attuned to sounds and other stimuli that would otherwise go unheeded. The more poorly you sleep, the more caffeine you tend to consume the next day, and the more lightly you sleep the follow-ing night. And so on. As Murray said upon seeing me refill my mug, "That's not a long-term solution, shipmate."

We're approaching periscope depth. The lights in the control room have been shut off. This is done for the benefit of the man at the periscope, who will shortly be taking a look around in the 5:00 a.m. darkness at the surface. To everyone else up here, many of whom are going on four or fewer hours of sleep, darkness is the opposite of helpful. Not only is it warm and dark in here, but because we're nearing the surface, the submarine is now rocking gently with the swells. "Torture," says the helmsman.

Torture was the word used by sleep researcher William Dement, who, as a student in the 1950s, helped Nathaniel Kleitman document

---

* To reduce troops' load, the Army adds caffeine to gum or mints or foods that soldiers are already carrying, like jerky. Natick public affairs officer David Accetta feeds a Caffeinated Meat Stick to reporters who visit the food lab. To me, it tasted just like you'd expect caffeinated meat to taste. Accetta was taken aback. "Brian Williams loved them." *Or did he?*

Rapid Eye Movement (REM) sleep. Before there were eyelid electrodes and electroocculargraphs, there were grad students pulling all-nighters. "Staring at the closed eyes of human adults by the dim light of a 30-watt bulb in the middle of the night was sheer torture," Dement wrote in a tribute to Kleitman, who is known in his field as "the Father of Sleep Research." (Tougher yet was the job of the chaperone Kleitman insisted be present when the subject was female: watching *someone else* watching the eyes of a sleeping human all night.)

THE PHOTOGRAPH dates from 1938. Nathaniel Kleitman sits at a dinner table, knife and fork crossed in a slab of hickory-smoked ham. What's unusual about this ham supper is that it took place in a cave 119 feet underground. Kleitman, with a graduate student assisting, spent thirty-two days in Kentucky's Mammoth Cave investigating the cycles of human sleep and wakefulness. He wished to find out: to what extent are these rhythms tied to external cues and routines? If you took away the cues—sunlight, established mealtimes, regular business hours—could people slip easily into an altered routine? Going underground seemed like the easiest* path to an answer.

---

* Though not, as correspondence in the Nathaniel Kleitman Papers reveals, without its challenges. To avoid "the danger of rats jumping up," the researchers' beds were outfitted with special five-foot-high legs with "tin rat guards." Alas, there was no rat guard for publicity-seeking tourist attraction managers and noisome reporters. Kleitman had made clear he wanted no press involvement, but about a week into the experiment, Mammoth Cave general manager W. W. Thompson sent a note down with the evening meal saying that reporters had *somehow, mysteriously,* found out about it and were clamoring for access. Kleitman did not go quietly. He asked to review the copy. He made *News of the Day* state in writing that they would "in no way ridicule the experiment." *Life* magazine got the last laugh: A "printer's mistake," the editors claimed in a letter of apology, caused Kleitman's title ("Dr.") to be "transposed" with the "Mr." before the name of his grad student.

Submarines interested Kleitman, because, like caves, they present a sort of real-world laboratory for chronobiology. Kleitman, in turn, interested the Submarine Force. They were, as they are today, having some alertness issues. Kleitman came up with a watch schedule that took advantage of a submarine's isolation from sunlight—the fact that it's always, as Murray put it, "70 degrees and fluorescent." It should thus be possible, Kleitman reasoned, to put each of three separate watch crews on a different schedule by staggering their waking hour, each crew beginning the day at a different time.

Beginning in 1949, three submarines, the *Corsair*, the *Toro*, and the *Tusk*, gave the Kleitman watchbill a two-week trial. At the end of it, Kleitman distributed questionnaires. "Should new schedule replace old?" read the last question. "Yes," said 19 sailors. "No," said 143. What happened? Catastrophic decompensation in the galley. Rather than cooking and cleaning up one breakfast, one lunch, and one dinner every twenty-four hours, galley crew had to do three of each, accommodating the different start times of each watch group's "day." The cooks were exhausted and peeved. The galley was a mess—"never clear and clean for more than an hour and a half," causing every meal to be "flavored with the odor of the last, and the whole permeated with the aura of aged refuse." And because a submarine's galley doubles as its rec room, movies could no longer be screened. "Recreational activities had to be curtailed to such an extent that they degenerated to periods of loafing around trying to keep out of the way." Friends who weren't assigned to the same "time zone" were now isolated from each other. "It is considered neither desirable nor feasible to continue this experimental watch schedule any further," concluded the final memo in the submarine folders of the Nathaniel Kleitman Papers.

There were some in the Submarine Force who believed Kleit-man's watchbill deserved a fuller chance, that the galley and rec-reation routines could have been adjusted. One executive officer blamed "just plain orneriness. Sailors," he wrote, "hate to try any-thing new." It is perhaps no coincidence that the colloquialisms "don't rock the boat" and "don't make waves" share a nautical element.

The officer was probably right. The Kleitman watchbill was grounded in sound science. Sunlight is our most powerful internal clock-setter. Along with rods and cones, we have a third kind of photoreceptor, one that is keyed to the blue wavelengths of sunlight. Information about this light, or the lack of it, is passed along to the pineal gland, producer of melatonin, the body's natural soporific. Sunlight triggers a cutoff of melatonin, bringing on wakefulness. (Indoor light—particularly the light from tablets and smartphones—can also suppress melatonin, but nowhere near as dramatically as sunlight.) This is why night shift workers who drive home in the morning through sunlight and then struggle to fall asleep may find relief by buying amber-lensed Bono-style glasses that block the sun's blue light wavelengths.

NSMRL has been developing goggles rimmed with battery-powered lights that emit the blue melatonin-suppressing wavelengths, thereby fooling the brain into thinking it's daytime. Depending on which direction you're flying, one or another of these distinctive eyewear options can help you preadjust to a new time zone. Or, in the case of Special Operations types heading to the Middle East to undertake secret 3:00 a.m. missions, *not* adjust. Lieutenant Kate Couturier, a circadian rhythm researcher at NSMRL, outfitted a planeload of Navy SEALs with blue light–emitting goggles on a series of flights from Guam to the East Coast of the United States, to see if it were possible to make them unattractive to females, oops, I mean, to keep them on Guam time. It worked.

It is probably fair to say that circadian dysrhythmia affects alertness and performance as much as or more than the amount of sleep a person has been getting. In the late 1990s, a team of sleep researchers and statisticians from Stanford University analyzed twenty-five seasons of *Monday Night Football* scores. Because the games were played at 9:00 p.m. eastern standard time, West Coast players were essentially competing at 6:00 p.m.—a time chronobiologically closer to the body's late afternoon peak for physical performance.* As the researchers predicted, West Coast teams were shown to have won more often and by more points per game. The effect was striking enough that teams sometimes travel a few days in advance of a game to give the players' body rhythms a chance to adjust.

Another complicating factor with military sleep is that the people making the schedules are often middle-aged and the people following them are teenagers. Not only do adolescents typically need more sleep, but their circadian rhythm is "phase-delayed" compared to adults'; melatonin production begins later in the evening, with the result that a teenager or even a twenty-two-year-old may not feel sleepy until well past midnight. Horrifically, the traditional boot camp lights-out is at 10:00 p.m., with a 4:00 a.m. wake-up.

Jeff Dyche told me about an admiral who approached him wanting to address sleep deprivation in the Navy boot camps. She wanted to move lights-out to an hour earlier—9:00 p.m.—to give the lads more time to sleep. Dyche quietly took her for a walk around camp after lights-out. "Almost every sailor was sitting up wide awake,

---

* It's not just alertness that waxes and wanes. Gut motility also follows a circadian pattern. Healthy humans rarely crap after midnight, unless they've just arrived in a distant time zone.

twiddling his thumbs. They're all going to sleep at midnight no matter how early they have to get up." Dyche managed to move some 4:00 a.m. wake-ups to 6:00 a.m. Test scores improved so dramatically that one of the command master chiefs assumed there'd been a cheating scandal.

For the past four decades, submarines have run on a watchbill known as "sixes," which divides sailors' time into six-hour chunks: six hours on watch, six for other duties and studies, six for personal time and sleep, then back on watch. The creation of an 18-hour day saw each sailor putting in six extra hours of watch time every 24-hour period. The problem is that his activities ceased to align with his biological rhythms. He's now working when his body badly wants to be sleeping. "It's like flying to Paris every day," Kate Couturier said. Without Paris. "It's a quadruple whammy," said Couturier's colleague Jerry Lamb, when I met with him and other Navy sleep experts before boarding the *Tennessee*. "We flip around their sleeping and working times, we work them like dogs, we give them very short periods of sleep, and we wake them up for drills." He turned to his colleagues. "Did I miss anything?"

Lamb was involved in the push for the new "circadian-friendly" watchbill. There has been, as there always is, some resistance. Sixes is how it's been done for fifty years. "As flawed as it is, we'd perfected it," commanding officer Bohner said one morning as we sat in his stateroom. "Now we're going to shake the ball and throw the pieces back down again." I tried to picture what game that might be.

The problem resides mainly with the midnight to 8:00 a.m. shift—the dreaded "mids." You come off watch and instead of sitting down to dinner, you're having breakfast. You're sleeping from 4:00 p.m. to 10:00 p.m., when there's often, despite Nathan Murray's best efforts, something that you have to get out of bed for. To more fairly distribute the suck, the crew swap watch schedules every other week. Instead of flying to Paris every day, it's every two weeks. The switchover happens on a Sunday, its being normally—that is, when riders are not coming aboard creating extra work for everyone—the quietest day of the week.

Today is that Sunday. Lieutenant Kedrowski, the man on the periscope platform, the officer of the deck, is switching to mids. It's his birthday. Happy birthday, Kedrowski. You get to scramble your circadian rhythms and get three hours sleep—in a rack that smells like someone else, because you had to give yours up to some writer from California.

"I'm really sorry, by the way." I would have been happy to sleep among the warheads.

"It's no problem," says Kedrowski, with unforced bonhomie. Almost everyone I've met down here has been easygoing and upbeat, especially given how tired they must be. I am, to quote the Dole banana carton in the galley pantry, "hanging with a cool bunch." If everyone in the world did a stint in the Navy, we wouldn't need a Navy.

Up above Kedrowski's head, a red light is flashing. Kedrowski explained this alarm box earlier. It's the one that goes off if the President of the United States orders a nuclear missile launch.

"So this is another drill then?"

"No." Kedrowski finishes writing something in a three-ring binder and looks over at the box. "It's kind of broken." He puts down his pen and listens. "They're supposed to say, 'Disregard alarm.'" They don't, and soon it stops. "They need to fix that," he says.

The missile alarm is mildly unnerving (good god, *what if?*) but not particularly frightening. In the queer logic of war in general and nuclear conflagration in specific, five hundred feet underwater on an undetectable Trident submarine is the safest place you could possibly be. The crew of the modern ballistic missile submarine endures long hours and grueling tedium, homesickness, horniness, and canned lima beans, but they are spared the thing that keeps most of us out of the military: the nagging awareness that you could be shot or blown up at almost any moment. Better dead-tired than dead.

14 Feedback from the Fallen

How the dead help the living stay that way

IT IS NOT THE blood in news photos of people shot dead or killed by bombs that gets to me. It's the clothing. Here's a man who got up in the morning and went to the closet with no inkling he was pulling on his socks for the last time, or adjusting his tie for the coroner. The clothing becomes a snapshot of a person's final, poignantly ordinary day on Earth. You see at once the death and the life. In autopsy photographs of US military dead, you also see what came between the two. Defense Department policy is to leave all life-saving equipment in place on a body. You see the urgent work of medics and surgeons—the pushing back at death with tourniquet and tube.

In military autopsies, medical hardware is examined alongside the software of organs and flesh. The idea is to provide feedback to the men and women who worked on these patients. Did, say, the new supraglottic airway device work the way the manufacturer promised? Was it placed correctly? Could anything have been done differently? The feedback happens via a monthly combat mortality teleconference, part of the Armed Forces Medical Examiner System (AFMES) program Feedback to the Field. In the past, solid, quantified feedback took the form of published papers. In the time it takes to have a study peer-reviewed and published in a medical journal, a lot of lives can be lost. This is so much better.

The System comprises two low tan brick buildings, mortuary and morgue. The mortuary being the one with the lovelier landscaping. Which is not to say that the morgue is bleak or depressing. It isn't (certainly not by comparison with what you have driven through to reach it: Dover Liquor Warehouse, Super 8 Motel, Chik-fil-A, Applebee's, Adult Probation and Parole, McDonald's, Wendy's, New Direction Addiction Treatment, Boston Market, and a giant blow-up rat advertising extermination services). A walkway connects the two buildings, but an ID badge is needed to open the door from one into the other. You don't want family members to take a wrong turn and end up in the autopsy suite.

Or, this morning, the conference room. The 7th Combat Mortality Conference is just getting under way. Eighty teleconnected individuals are taking part: thirty or so here at AFMES, about an equal number phoning in from Afghanistan and Iraq, and a few in San Antonio, Texas, at the US Army Institute of Surgical Research. They interact by audio only. There is a video screen, but it is used to display not the speakers but the soldiers being spoken about.

The body in the photograph on the screen is on its back. Black bars have been added over the eyes and groin. I want a third one to hide the feet, which are flopped strangely, wrongly, off to the same side. They're like feet in an ancient Egyptian frieze or under the bedding at one of those hotels where the maids aggressively tuck the sheets. A man speaking from Afghanistan recites the prehospital care scenario. "CPR was in progress when he arrived. Treatment included JETT tourniquet, sternal intraosseous IV, plasma, two doses epinephrine. Upon arrival at the medical treatment facility, no cardiac activity was noted. CPR was ceased. Over." *Over* is of course the military man's habit of denoting the end of a radio communication, not, as I at first heard it, a dramatic editorial flourish.

The medical examiner (or ME) who performed the autopsy delivers his summary next. ". . . Extensive head injuries, skull fractures. Laceration of the brain stem. Hemorrhages. Multiple facial fractures. Extensive injuries to the upper extremities. Also fractures of both his tibiae and fibulae. Facially again, his maxilla and mandible are fractured." The blood has been cleaned away, so most of what I'm hearing doesn't register visually. What registers is this: *His mustache is on crooked.* It calls to mind the old slapstick gag—the false mustache slipping its glue and hanging askew on the actor's face. It wasn't that funny then, and it's very much not so now.

Like a sensitive editor, the medical examiner begins his feedback with positives. "Crike was performed adequately and perfectly placed." Crike is short for cricothyrotomy—puncturing an emergency airway through the cricothyroid membrane. The ME moves on. "Placement of JETT." The Junctional Emergency Treatment Tool is a new type of tourniquet for compressing the femoral arteries at the junction of the leg and the torso. "JETT was possibly moved in transport . . ." This is polite shorthand for the fact that it is not where it should be. Great care is taken at these meetings with how things are phrased. The medical examiners don't want to lay blame or criticize the people who provided the care. Rather than refer to them by name or classification, they say "the user of the device."

A man at the army surgical research institute has something to add. "It's a rookie mistake," he begins, "to place the JETT too proximal. The femoral artery is more easily compressed slightly distal"—farther out—"to where this device was placed." He backtracks. "Though it's possible the device moved proximally in transit." He can't help adding: "Though it's not likely that happened. I'll send the instructions to everyone. Thanks."

The next case is easier to look at, too easy. There shouldn't be a situation in which you find yourself admiring the build of a dead person. When all is right with the world, corpses look ancient, weak, worn out. You know at a glance there was little living left to be done with that body. "You can see the sternal IO has been placed properly," the ME is saying. IO stands for intraosseous. It's the cousin of IV—intravenous. IO refers to a blood transfusion via the bone marrow rather than a vein. When someone has lost a large volume of blood, the vessel walls lose the tautness that makes it possible to find them and pierce them with a needle. It's the difference between poking a pin into a newly inflated balloon and one that's kicking around in the corner a week after the party. The bone—often the sternum, which puts out a lot of blood—is breached via a small drill or gun, or a determined twisting by hand if the batteries are low.

In days past, this man's glorious pectorals could have been a party to his demise. One of the pieces of feedback AFMES gave to the field was this: The pectoral muscles of the modern, weight-lifting soldier or Marine are often so bulked up that the needle inserted into the chest to relieve air pressure in cases of collapsed lung—if, say, the lung is shot through and air is building up outside of it—isn't long enough to clear the muscles. This was the case in about half of all male patients. Because of Feedback to the Field, longer needles are used on buffer soldiers.

The last case is a woman shot from behind. The ME is narrating. "The larger of two penetrating gunshot wounds crosses through the heart into the right lung. . . . Proper location of sternal IO. Proper location of tibial IO." There isn't much else to say. There wasn't much else to be done.

The woman's underwear has been left on. It is pale yellow* and plain. The image guts me. It's the clothing thing—the innocence of the unsuspecting. In a second slide, the body lies facedown. The back side of the underwear, I notice, is pink. It takes a moment to process why. Yellow plus blood equals pink.

THE AUTOPSY room smells like summer. The exhaust system has an air intake from outside, explains AFMES public affairs officer Paul Stone, who is taking me around this afternoon. "They just mowed the grass out there." The room is large enough to accommodate twenty-two autopsies at once. Stone was here the week a Chinook helicopter was shot down in Afghanistan, killing thirty-eight people and a military working dog. Then, it smelled like jet fuel and burned flesh, so powerfully that Stone's dry cleaner charged him double. "He said, 'What were you *doing?*'" Stone used to be a spokesperson at the Office of the Secretary of Defense. It's tough

---

*Female soldiers, unlike males, receive vouchers to shop for their own underthings. The US military is gearing up to buy uniforms embedded with photovoltaic panels—*shirts that can recharge a radio battery*—but it is not up to the task of purchasing bras for female soldiers. "I've done that sort of shopping with my wife," said an Army spokesman quoted in *Bloomberg Business*. "It's not easy to do."

to rattle him. At one point I asked if people tell him he looks like Vladimir Putin, and not even that did it.

At the peak of the Iraq war, twenty or thirty bodies passed through this room each week. Since 2004, around six thousand autopsies have taken place here. Every person (and dog) who dies in the service of the US military is autopsied. It was not always this way. Before 2001, autopsies were reserved for cases in which there was no witness to the death, or the cause was not obvious. Stone gives the example of a suspected homicide, then pauses. "Though technically it's all homicide." *Homicide*, from the Latin *homo*, for man, and *-cidium*, the act of killing. He means murder: prosecutable homicide.

Six thousand *homos cidium*ed in the prime of their lives. What does this job do to a person? For one thing, it makes him very tired of that question. "We're doctors, and these are our patients," was the stock answer I got. I imagine it's a tough kind of doctor to be. Most people study medicine with the hope and intent that their work will restore health, end pain, extend lives. *Save* lives. Because of Feedback to the Field, the work of these medical examiners does save lives. But not the ones they interact with day to day.

Stone brings me over to the H. T. Harcke Radiology Suite, where dead men and women are given CT scans. A whole-body CT is a heavy dose of radiation, but the dead don't have to worry. Certain things like bullet trajectories and angles of entry are easier to see in the clean, gray-scale imagery of a CT scan than they are in a flesh-and-blood autopsy. Colonel Harcke himself is on hand to show me the basics of forensic radiopathology. He is the Harcke for whom the lab is named. I assumed that this was in tribute to his pioneering contributions to the field. "There's two ways that happens," he says when I mention it. "Die or give two million dollars. I'll let you figure out which it is."

Using a mouse, Harcke scrolls through the topography of an anonymous body. As we travel from scalp to boot heel, IED fragments flare like supernovae. Metal reads as bright white against the grays of muscle, blood, and bone.* The contrast is stark and telling. In the face of velocitized steel, even the strongest among us are mush. Fragility is evident even in the terms MEs use—*soft tissue*, an *eggshelled skull*.

On the way back to Stone's office, we stop to talk with Pete Seguin, the statistics guy. On his desk is a sheaf of photographs, printouts of the cases from the combat mortality meeting. "They don't look real," he says of the bodies. "They're like dolls." I'm not sure where he's buying his dolls. I look at Stone.

"He means porcelain dolls," Stone says. "The white skin." Seguin explains lividity, the pooling of blood in a corpse. When the pump shuts down, gravity takes over. Because the dead are transported on their backs, they come to autopsy white as geishas, the blood drained from the face, chest, the tops of the legs.

"But then you see them back there . . ." Seguin means in the autopsy room. "That's a whole different experience. It's too sad." I can barely hear him. "These are all young people. Our kids. It makes you ask questions. Like, Was it worth it?"

In the autopsy room there's a pair of platformed aluminum stepladders on wheels. I thought the ceiling was being repaired. "No, it's

---

* Usually the victim's, but occasionally a fragment from a suicide bomber. According to Stone, there has not been a documented case in which a piece of a terrorist's bone was the cause of death. (Medical examiners do not use the term "organic shrapnel." That originated in *Falling Man* author Don DeLillo's cranium.)

for perspective," Stone had said. The autopsy photographers need to get up high to get the whole body in the frame. I guess war is like that. A thousand points of light, as they say. Only when you step back and view the sum, only then are you able to grasp the worth, the justification for the extinguishing of any single point. Right at the moment, it's tough to get that perspective. It's tough to imagine a stepladder high enough.

Acknowledgments

This book began with an email from a reader: Brad Harper, a retired Army pathologist. In the course of our correspondence, I mentioned I'd been toying with the idea of a book on military science but had assumed that access would trip me up. Should I try it anyway? Yes, insisted Harper. He brought me to the military morgue in Dover and introduced me to colleagues. He took me to USUHS to see his friend Sharon Holland, who has contacts all over the military medical world. When I allowed that one of the things I wished to write about was genital trauma, Holland did not flinch. She picked up the phone and called James Jezior at Walter Reed. *Hey, Jim, might you have a surgery this writer could observe?* Yes, said Jezior. Though he'd need to ask the patient. And surely here would be my first no: *Hey, Captain White, could some strange writer lady come out and watch your operation?* But White, too, said yes.

And so it went. Over and over, when the easy answer, the sane answer, was no, people said yes.

Hey, Jerry Lamb, ridiculously busy technical director at the

Naval Submarine Medical Research Laboratory, could you find someone to approve my spending a few days at sea on a Trident submarine? Though it'll take fourteen months and two-hundred-some emails to make it happen? Yes, said Lamb.

And might that submarine be yours, Chris Bohner and Nathan Murray of the USS *Tennessee?* Though I'll be traipsing through the missile silos with no security clearance? Yes, they said. Bring your notebook and your dingbat questions. Kick Kedrowski out of his rack. Tie up the head every morning.

Hello, Mark Riddle, could I follow you to Camp Lemonnier, Djibouti, even though it means you'll have to escort me all day every day for an entire week? And then later will you spend your holidays reviewing my manuscript?

Hey, Randy Coates, and hey, Rick Redett, I hear you're doing some cadaver trials. Could I join you?

Hey, Kit Lavell, hey, Eric Fallon, could you work me into combat simulations where I don't belong?

Again and again, I expected to hear no, yet yes was what I got. These fine people put their reputations on the line. They spent time they could not spare. They spoke openly on issues more comfortably left alone. For all of this—to all of you—I am deeply, humbly, gobsmackedly grateful.

I have no background in medicine or the military, and this fact made me an exasperating, time-sucking presence in people's days. Certain individuals must be thanked for the hours spent explaining their work and, in some cases, the most basic elements of the science: Rob Dean, Christine DesLauriers, Molly Williams, Benjamin Potter, and Stacy and Mark Fidler at Walter Reed National Military Medical Center; Doug Brungart, Ben Sheffield, George Peck, Dan Szumlas, and Pete Weina at Walter Reed Army Institute of Research; Natalie

Pomerantz, Sam Cheuvront, Peggy Auerbach, Rick Stevenson, and Annette LaFleur at US Army Natick Soldier Research, Development and Engineering Center; Alan Hough and Eric Nabors at the US Navy Submarine School; Kate Couturier, Ray Woolrich, and Shawn Soutiere at Naval Submarine Medical Research Laboratory; Dianna Purvis, Patty Deuster, and Dale Smith at Uniformed Services University of the Health Sciences; Mark Roman at Aberdeen Proving Ground; Ken Tarcza, Jason Tice, and Patti Rippa of the Warrior Injury Assessment Manikin project; Nicole Brockhoff in the Office of the Director, Operational Test and Evaluation; Aaron Hall and Dave Regis at the Naval Medical Research Center; Theodore Harcke and Edward Mazuchowski at the Armed Forces Medical Examiner System; John Clark of the Royal Navy, and Michael White. I came to you all as an ignoramus and an outsider, and you treated me as neither.

Outside the military, I made a pest of myself, most notably, with David Armstrong, Charlie Beadling, Greg Belenky, John Bolte, Robert Cantrell, Joe Conlon, Damon Cooney, Pam Dalton, Jeff Dyche, Jerry Hogsette, Andrew Karam, Malcolm Kelley, Darren Malinoski, Chris Maute, Ekaterina Pesheva, Bruce Siddle, Terry Sunday, and Ronn Wade. I am grateful for the patience and unflagging good humor accorded to me by all.

The stereotype of the military spokesperson—the obfuscating spin-doctor who prefers to pass the buck—was nowhere in evidence during the writing of this book. The public affairs people I contacted were accommodating and no-bull. A few stand out for the extreme diligence and tolerance they applied to my off-the-grid inquiries: David Accetta at US Army Natick Soldier Research, Development and Engineering Center; Seamus Nelson at US Navy Camp Lemonnier; Dora Lockwood at the Navy Bureau of Medicine and Surgery;

Doris Ryan of the Naval Medical Research Center; Paul Stone at the Armed Forces Medical Examiner System; Joyce Conant at US Army Research Laboratory; Joe Ferrare at US Army Research, Development and Engineering Command; and Jenn Elzea and Sue Gough at the Office of the Secretary of Defense.

Vast troves of military images and archival material are available to those who know where and how to look. I knew neither. I am beholden to Andre Sobocinski, able historian at the US Navy Bureau of Medicine and Surgery, for helping me navigate the National Archives and Records Administration and cheerfully photocopying entire folders for me. Likewise, I owe an outsized debt of thanks to Stephanie Romeo for chasing down the images that open each chapter. Her generous nature and zeal for the task led her to spend far more time than she had any good reason to spend.

Yet again, I have benefited from the unerring instincts of some extraordinary people in the publishing world. I am so very lucky to have Jill Bialosky as an editor and friend. Erin Lovett and Louise Brockett, Bill Rusin, Jeannie Luciano, Drake McFeely, Ingsu Liu, Steve Colca, Laura Goldin, and Maria Rogers at W. W. Norton make my job a joy and my books the best they can be. Jay Mandel's support is the bass line that runs through my career, and Janet Byrne is the best copy editor imaginable.

No matter how well things fall into place and how smoothly the writing goes, a book will send you to the couch in occasional fits of doubt and self-pity. Everlasting love and gratitude to my husband, Ed Rachles, the man who gets me off the couch.

Bibliography

BY WAY OF INTRODUCTION

Beason, Robert C. "What Can Birds Hear?" *USDA National Wildlife Research Center—Staff Publications*, Paper 78, 2004.

Lethbridge, David. "'The Blood Fights on in Other Veins': Norman Bethune and the Transfusion of Cadaver Blood in the Spanish Civil War." *Canadian Bulletin of Medical History* 29, no. 1 (2012): 69–81.

Speelman, R. J. III, M. E. Kelley, R. E. McCarty, and J. J. Short. "Aircraft Birdstrikes: Preventing and Tolerating." IBSC-24/WP31. Wright-Patterson Air Force Base, OH: Air Force Research Laboratory, 1998.

1. SECOND SKIN

"DOD Should Improve Development of Camouflage Uniforms and Enhance Collaboration Among the Services." Report to

Congressional Requesters. Washington, DC: US Government Accountability Office, 2012.

Oesterling, Fred. "Thermal Radiation Protection Afforded Test Animals by Fabric Ensembles." Operation Upshot-Knothole Project 8.5: Report to the Test Director. WT-770. Quartermaster Research and Development Laboratories, 1955.

Phalen, James M. "An Experiment with Orange-Red Underwear." *Philippine Journal of Science* 5, no. 6 (1910): 525–46.

White, Bob. "How Your Meat Helps Your Men." *Breeder's Gazette*, July–August 1943, 20–21.

2. BOOM BOX

Balazs, George C., et al. "High Seas to High Explosives: The Evolution of Calcaneus Fracture Management in the Military." *Military Medicine* 179, no. 11 (2014): 1228–35.

Warrior Injury Assessment Manikin (WIAMan) Project Boot Fitting Procedures, version 1.2. Warrior Injury Assessment Project Management Office: November 10, 2015. Distribution Statement, W0060.

3. FIGHTING BY EAR

Berger, Elliott H. "History and Development of the E-A-R Foam Earplug." *Canadian Hearing Report* 5, no. 1 (2010): 28–34.

Bradley, J. Peter. "An Exploratory Study on Sniper Well-Being." Defence R&D Canada–Toronto, Contractors Report. DRDC Toronto CR 2009-196. 2010.

McIlwain, D. Scott, Kathy Gates, and Donald Ciliax. "Heritage of Army Audiology and the Road Ahead: The Army Hearing

Program." *American Journal of Public Health* 98, no. 12 (2008): 2167–72.

Sheffield, Benjamin, et al. "The Relationship Between Hearing Acuity and Operational Performance in Dismounted Combat." *Proceedings of the Human Factors and Ergonomics Society Annual Meeting* 59, no. 1 (2015): 1346–50.

4. BELOW THE BELT

Dismounted Complex Injury Task Force. "Report of the Army: Dismounted Complex Blast Injury." June 18, 2011.

Ellis, Kathryn, and Caitlin Dennison. *Sex and Intimacy for Wounded Veterans: A Guide to Embracing Change*. The Sager Group, 2014.

5. IT COULD GET WEIRD

Dubernard, Jean-Michel. "Penile Transplantation?" *European Urology* 50 (2006): 664–65.

Hu, Weilie, et al. "A Preliminary Report of Penile Transplantation: Part 2." *European Urology* 50 (2006): 1115–16.

Reed, C. S. "The Codpiece: Social Fashion or Medical Need?" *Internal Medicine Journal* 34 (2004): 684–86.

6. CARNAGE UNDER FIRE

Arora, Sonal, et al. "The Impact of Stress on Surgical Performance: A Systematic Review of the Literature." *Surgery* 147, no. 3 (2009): 318–30.

Landis, Carney, William A. Hunt, and Hans Strauss. *The Startle Pattern*. New York: Farrar & Rinehart, Inc., 1939.

Love, Ricardo M. *Psychological Resilience: Preparing Our Soldiers for War.* Carlisle Barracks, PA: US Army War College, 2011.

Webb, Brandon. "A Kit Up Inside Look at 'Goat Lab.'" February 21, 2012. http://kitup.military.com/2012/02/goat-lab-an-inside-look.html.

7. SWEATING BULLETS

Adolph, E. F. *Physiology of Man in the Desert.* New York: Interscience Publishers, 1947.

Carter, Robert III, et al. "Epidemiology of Hospitalizations and Deaths from Heat Illness in Soldiers." *Medicine and Science in Sports and Exercise* 37, no. 8 (2005): 1338–44.

Heat Injuries, Active Component, U.S. Armed Forces. *Medical Surveillance Monthly Report* 19, no. 3 (2011): 14–16.

Kuno, Yas. *Human Perspiration.* Springfield, IL: Charles C. Thomas, 1956.

Tucker, Patrick. "The Very Real Future of Iron Man Suits in the Navy." Defense One, January 12, 2015. www.defenseone.com/technology/2015/01/very-real-future-iron-man-suits-navy.102630.

Update: Exertional Rhabdomyolysis, Active Component, US Armed Forces, 2011. *Medical Surveillance Monthly Report* 19, no. 3 (2012): 17–19.

8. LEAKY SEALS

Barbeito, Manuel S., Charles T. Mathews, and Larry A. Taylor. "Microbiological Laboratory Hazard of Bearded Men." Technical Manuscript 379. Frederick, MD: Department of the Army, 1967.

Connor, Patrick, et al. "Diarrhoea During Military Deployment:

Current Concepts and Future Directions." *Journal of Infectious Diseases* 25, no. 5 (2012): 546–54.

Dandoy, Suzanne. "The Diarrhea of Travelers: Incidence in Foreign Students in the United States." *California Medicine* 104, no. 6 (1966): 458–62.

Lim, Matthew L., et al. "History of US Military Contributions to the Study of Diarrheal Diseases." *Military Medicine* 170, no. 4 (2005): 30–38.

Porter, Chad K., Nadia Thura, and Mark S. Riddle. "Quantifying the Incidence and Burden of Postinfectious Enteric Sequelae." *Military Medicine* 178, no. 4 (2013): 452–59.

Sanders, John W., et al. "Impact of Illness and Non-Combat Injury During Operations Iraqi Freedom and Enduring Freedom." *American Journal of Tropical Medicine and Hygiene* 73, no. 4 (2005): 713–19.

Vaughan, Victor. "Conclusions Reached After a Study of Typhoid Fever Among the American Soldiers in 1898." *Journal of the American Medical Association* 34 (June 9, 1900): 1451–59.

9. THE MAGGOT PARADOX

Baer, William S. "The Treatment of Chronic Osteomyelitis with the Maggot (Larva of the Blow Fly)." *Journal of Bone and Joint Surgery* 13, no. 3 (1931): 438–75.

Fennell, Jonathan. *Combat and Morale in the North African Campaign: The Eighth Army and the Path to El Alamein.* Cambridge, UK: Cambridge University Press, 2014.

Filth Flies: Significance, Surveillance and Control in Contingency Operations. Armed Forces Pest Management Board Technical Guide No. 30. Washington, DC: Armed Forces Pest Management Board Information Services Division, 2011.

Heitkamp, Rae A., George W. Peck, and Benjamin C. Kirkup. "Maggot Debridement Therapy in Modern Army Medicine: Perceptions and Prevalence." *Military Medicine* 177, no. 11 (2012): 1411–15.

Kenney, Michael. "Experimental Intestinal Myiasis in Man." *Proceedings of the Society for Experimental Biology and Medicine* 60 (November 1945): 235–37.

Lenhard, Raymond. *William Stevenson Baer: A Monograph*. Baltimore: Schneidereith & Sons, 1973.

Lovell, Stanley. *Of Spies and Stratagems*. Englewood Cliffs, NJ: Prentice-Hall, 1963.

Miller, Gary L., and Peter H. Adler. "' . . . Drenched in Dirt and Drowned in Abominations . . . ': Insects and the Civil War." In *Proceedings of the DOD Symposium on Evolution of Military Medical Entomology*, November 16, 2008.

Sharpe, D. S. "An Unusual Case of Intestinal Myiasis." *British Medical Journal*, January 11, 1947, 54.

Sherman, R. A., M. J. R. Hall, and S. Thomas. "Medicinal Maggots: An Ancient Remedy for Some Contemporary Afflictions." *Annual Review of Entomology* 45 (2000): 55–81.

10. WHAT DOESN'T KILL YOU WILL MAKE YOU REEK

Washington Services Branch Records. Record Group 226: Records of the Office of Strategic Services. 1919–2002.

11. OLD CHUM

Baldridge, David H., Jr. "Analytic Indication of the Impracticability of Incapacitating an Attacking Shark by Exposure to Waterborne Drugs." *Military Medicine* 134 (November 1969): 1450–53.

—————. "Shark Attack: A Program of Data Reduction and Analysis." Published as a monograph entitled *Contributions from the Mote Marine Laboratory*, Volume 1, Number 2, 1974.

Baldridge, David H., Jr., and L. J. Reber. "Reaction of Sharks to a Mammal in Distress." *Military Medicine* 131, no. 5 (May 1966): 440–46.

Castro, José I. "Historical Knowledge of Sharks: Ancient Science, Earliest American Encounters, and American Science, Fisheries, and Utilization." *Marine Fisheries Review* 75, no. 4 (2013): 12–25.

Cushing, Bruce S. "Responses of Polar Bears to Human Menstrual Odors." *International Conference on Bear Research and Management* 5 (1980): 270–74.

Golden, Frank, and Michael Tipton. *Essentials of Sea Survival*. Champaign, IL: Human Kinetics, 2002.

Llano, George A. "Open-Ocean Shark Attacks." Chapter in *Sharks and Survival*, edited by Perry Gilbert. Boston: D. C. Heath, 1963.

Rogers, Lynn, Gregory A. Walker, and Sally S. Scott. "Reactions of Black Bears to Human Menstrual Odors." *Journal of Wildlife Management* 55, no. 4 (1991): 632–34.

Tester, Albert. "The Role of Olfaction in Shark Predation." *Pacific Science* 27 (April 1963): 145–70.

12. THAT SINKING FEELING

"Loss of the USS *Tang*." In *Medical Study of the Experiences of Submariners as Recorded in 1,471 Submarine Patrol Reports in World War II*, edited by Ivan F. Duff. Washington, DC: US Navy Bureau of Medicine and Surgery, 1960.

Giersten, J. C., et al. "An Explosive Decompression Accident." *American Journal of Forensic Medicine and Pathology* 9, no. 2 (1988): 94–101.

Karam, Andrew. *Rig Ship for Ultra Quiet: Living and Working on a Nuclear Submarine at the End of the Cold War.* Hartwell, Australia: Temple House, 2002.

Maas, Peter. *The Rescuer.* Alternate title for *The Terrible Hours.* New York: Harper & Row, 1967.

13. UP AND UNDER

Dement, W. C. "Remembering Nathaniel Kleitman." *Archives Italiennes de Biologie* 139 (2001): 11–17.

Friedl, Karl E. "Medical Technology Repurposed to Enhance Human Performance." In Overview of the HFM-Symposium Programme, October 5, 2009. RTO-MP-HFM-181.

Mackowiak, Philip A., Frederic T. Billings III, and Steven S. Wasserman. "Sleepless Vigilance: 'Stonewall' Jackson and the Duty Hours Controversy." *American Journal of the Medical Sciences* 343, no. 2 (2012): 146–49.

Miller, Nita Lewis, Lawrence G. Shattuck, and Panagiotis Matsangas. "Sleep and Fatigue Issues in Continuous Operations: A Survey of U.S. Army Officers." *Behavioral Sleep Medicine* 9 (2011): 53–65.

Rattenborg, N.C., S. L. Lima, and C. J. Amlaner. "Facultative Control of Avian Unihemispheric Sleep Under the Risk of Predation." *Behavioral Brain Research* 105, no. 2 (November 15, 1999): 163–72.

"Special Crew Rest Edition." CSL-CSP Force Operational Notes Newsletter. N.d. http://my.nps.edu/documents/105475179/105675443/FON+Newsletter+Sleep+Edition+-+Final.pdf/66e0b291-3708-428b-844d-e633d6c50527.

Smith, Roger S., Christian Guilleminault, and Bradley Efron. "Circadian Rhythms and Enhanced Athletic Performance in the National Football League." *Sleep* 20, no. 5 (1997): 362–65.

14. FEEDBACK FROM THE FALLEN

Eastridge, Brian J., et al. "Death on the Battlefield (2001–2011): Implications for the Future of Combat Casualty Care." *Journal of Trauma and Acute Care Surgery* 73, no. 6 (December 2012): Supplement 5, S431–S437.